Geostatistical Insights on Mapping Flood Hazards and Wetland Dynamics

Walid Chouari
College of Arts, King Faisal University, Saudi Arabia

Abdelkader Mohamed Sghaier Derbali
Taibah University, Saudi Arabia

IGI Global
Scientific Publishing
Publishing Tomorrow's Research Today

Vice President of Editorial — Melissa Wagner
Director of Acquisitions — Mikaela Felty
Director of Book Development — Jocelynn Hessler
Production Manager — Mike Brehm
Cover Design — Jose Rosado

Published in the United States of America by
IGI Global Scientific Publishing
701 East Chocolate Avenue
Hershey, PA, 17033, USA
Tel: 717-533-8845
Fax: 717-533-7115
Website: https://www.igi-global.com E-mail: cust@igi-global.com

Library of Congress Cataloging-in-Publication Data

LCCN: 2025020969 (CIP Data Pending)
ISBN13: 9798337332062
Isbn13Softcover: 9798337332079
EISBN13: 9798337332086

British Cataloguing in Publication Data
A Cataloguing in Publication record for this book is available from the British Library.

Table of Contents

Detailed Table of Contents

This study presents an analysis of flood hazard mapping in the Sub-Saharan African region and its impact on sustainable development. Flooding remains a significant challenge in the region, exacerbated by climate change, rapid urbanization, and inadequate infrastructure. Effective flood hazard mapping, using geographic information systems (GIS), remote sensing, and machine learning, is crucial for disaster preparedness, risk mitigation, and urban planning. However, limited access to accurate data, insufficient technical expertise, and funding constraints hinder the development of comprehensive flood mapping systems. The resulting vulnerabilities lead to displacement, economic losses, and disruptions in agriculture, directly affecting food security and livelihoods. Addressing these challenges requires a multi-stakeholder approach, integrating policy reforms, community engagement, and technological advancements. This study explores the role of flood hazard mapping in sustainable development and climate-resilient infrastructure to mitigate flood risks and enhance regional sustainability.

The study focuses on purchase and usage behavior towards arecanut products using the theory of planned behavior. As we are aware, arecanut is a traditional product consumed by people in India. The objective of the study is to identify the factors affecting purchase and usage behavior towards arecanut products and the effect of purchase attitude, subjective norm, perceived behavior control, intention to buy, satisfaction, and facilities on purchase and usage behavior. The research is exploratory in nature because it tries to explore purchase and usage behavior towards arecanut products. The study shows factors such as purchase attitude, subjective

Chapter 5

Ananya Pandey, Christ University, Bangalore, India
Achyutananda Mishra, Christ University, Bangalore, India

This chapter explores the evolving landscape of Farmers' Rights in the context of intensifying agricultural intellectual property regimes. It critically examines the intersection of international legal frameworks, national legislations, and customary practices that shape farmers' access to seeds, biodiversity, and knowledge systems. Through a multidisciplinary lens combining legal theory, environmental justice, and agroecological insights, the chapter interrogates how intellectual property regimes, such as UPOV and TRIPS, impact traditional seed systems, biodiversity conservation, and food sovereignty. It highlights tensions between privatized biotechnological innovations and community-centric stewardship of genetic resources, particularly in the face of climate change and digital sequence information. By tracing the trajectory from soil to statute, farmers' rights emerge from grassroots agricultural practices and struggles, evolving into legal recognition and protection through policy and law.

Chapter 6

Ananya Pandey, Christ University, Bangalore, India

This chapter explores the transformative potential of artificial intelligence (AI) in mitigating environmental hazards and enhancing resilience to climate-related challenges. AI technologies are revolutionizing environmental monitoring, risk assessment, and disaster response through advanced data analytics, machine learning, and predictive modeling. By leveraging AI, stakeholders can predict and mitigate the impacts of natural disasters, improve resource management, and foster resilient infrastructures. The chapter examines the various applications of AI in hazard prediction, such as flood forecasting, wildfire management, and storm tracking, alongside its role in real-time monitoring and adaptive response systems. Furthermore, it highlights the ethical considerations, challenges, and limitations of deploying AI in environmental management. Through case studies and emerging practices, this chapter outlines pathways for harnessing AI to safeguard communities and ecosystems, ensuring a more resilient and sustainable future.

The purpose of studying the nexus between urban planning and sustainability through a bibliometric perspective is to systematically analyze and understand the existing body of scholarly literature on this topic. This paper aims to review articles related to urban planning and sustainability with the help of bibliometric analysis. A total number of 899 documents were drawn from the Scopus database out of which 377 articles were selected. To accomplish the overview of Urban planning and sustainability Research, bibliometric techniques such as performance analysis and scientific mapping were applied. The data was also organized, analyzed, and presented using the Bibliometrix package. The study found that research on Urban planning and sustainability Research has a significant growth rate of -15.39 percent. It was also discovered that DESLATTE A has produced the most influential articles in this field and Geojournal was the top-ranked journal. The findings also highlighted that USA, India and China had the highest rates of research productivity in this field.

The Sundarbans region is home to diverse tribal communities facing significant health challenges exacerbated by flood hazards and wetland dynamics. Bridging health and habitat, this study explores the right to health for Sundarbans tribals through geostatistical insights. The correlation between environmental factors and health outcomes is critical, as these communities rely on wetlands for their livelihoods. Flooding not only disrupts access to essential resources but also increases vulnerability to waterborne diseases. By employing geostatistical methods, researchers can map flood-prone areas and wetland changes, facilitating proactive health interventions. This holistic approach underscores the need for policies that integrate environmental sustainability with healthcare access, ensuring that tribal populations can secure their right to health amidst changing climate conditions.

Addressing these intertwined issues is vital for enhancing resilience and improving overall well-being in the Sundarbans.

Preface

INTRODUCTION

We are pleased to present *Geostatistical Insights on Mapping Flood Hazards and Wetland Dynamics*, a comprehensive volume that brings together cutting-edge research and practical applications on two of the most pressing environmental challenges of our time: flood risk and wetland management. This book was conceptualized with a clear objective — to offer a geostatistical lens through which scientists, planners, policymakers, and stakeholders can better understand and address the spatial complexities of flood hazards and the dynamic nature of wetlands.

Flood mapping plays a vital role in contemporary environmental planning and disaster risk reduction. It is not merely a scientific or cartographic exercise; it is a tool of empowerment. The ability to inform current and future property owners and tenants of their exposure to flood risks enables individuals and communities to make informed decisions, safeguard lives, protect assets, and invest in resilience. Accurate flood maps guide mitigation strategies, influence insurance frameworks, and shape zoning and infrastructure development.

Equally significant are wetlands—marshes, swamps, and other water-saturated lands, both natural and artificial. Often undervalued, these ecosystems are rich in biodiversity and perform critical functions such as water purification, flood attenuation, and carbon sequestration. Whether they are seasonally inundated or permanently waterlogged, saline or freshwater, wetlands support agriculture, fisheries, and cultural heritage while also acting as buffers against environmental shocks. Recognizing their importance, this book gives wetlands the analytical attention they deserve within the broader scope of hydrological and land-use studies.

This volume compiles contributions from experts across disciplines and geographies, providing a global perspective on the technological, methodological, and policy-driven innovations that inform flood hazard and wetland mapping. Through the lens of geostatistics, we examine the power of spatial analysis, data integration, remote sensing, and predictive modeling to inform environmental management.

We hope this book will serve not only as a reference for researchers and practitioners but also as a catalyst for interdisciplinary dialogue and action. The challenges posed by climate variability, urbanization, and ecosystem degradation demand sophisticated tools and collaborative efforts. It is our belief that *Geostatistical Insights on Mapping Flood Hazards and Wetland Dynamics* will contribute meaningfully to these ongoing efforts and inspire further research and innovation.

ORGANIZATION OF THE BOOK

This book begins with a regional lens in **Chapter 1**, *Mapping Flood Hazards in Sub-Saharan African Region: Hindering Regional Sustainable Development*, where the authors explore the complex interplay between inadequate flood hazard mapping and sustainable development in Sub-Saharan Africa. Through geospatial tools such as GIS, remote sensing, and machine learning, the chapter highlights the urgent need for enhanced mapping practices in the face of climate change and rapid urbanization. A lack of accurate data and infrastructure poses significant risks to food security and livelihoods, necessitating integrative approaches involving community engagement, policy innovation, and technological interventions.

In **Chapter 2**, *Examining Purchase and Usage Behavior Towards Arecanut Products Using Theory of Planned Behavior*, the authors adopt a behavioral lens to investigate consumer interactions with arecanut products, a culturally significant commodity in India. Applying the Theory of Planned Behavior, the chapter identifies key variables such as purchase attitude, perceived behavioral control, and intention to buy that influence consumer decisions. Although distinct from flood mapping, this chapter broadens the book's interdisciplinary scope, offering a behavioral science methodology that can be translated into environmental consumption and policy behaviors.

Chapter 3, *Artificial Intelligence and Machine Learning for Geodata Analysis*, delves into the transformative applications of AI and ML in geospatial research. The chapter emphasizes the power of deep learning models—particularly convolutional neural networks—for land use classification and environmental monitoring. These technologies enable a more accurate, efficient analysis of large geospatial datasets, thus enhancing decision-making in flood risk mapping, disaster evaluation, and climate resilience efforts.

Chapter 4, *Approaches in Smart Urban Planning with Artificial Intelligence Through Data Science*, continues the data-driven narrative by addressing how AI and data science reshape urban sustainability and smart city planning. The authors showcase various applications where predictive analytics and algorithmic modeling are being utilized to design sustainable urban infrastructures. This chapter situates flood hazard and wetland dynamics within broader urban ecosystems, emphasizing the need for AI-integrated planning to achieve long-term resilience.

The discourse turns legal and ethical in **Chapter 5**, *Farmers' Rights in the Age of Agricultural Intellectual Property: From Soil to Statute*. This chapter examines the legal dimensions of agricultural innovation, highlighting tensions between international property regimes and indigenous seed sovereignty. With implications for biodiversity and traditional ecological knowledge, the discussion reaffirms the interconnectedness of land rights, resource access, and environmental justice, which are essential in managing wetland ecosystems and climate resilience.

Chapter 6, *Harnessing AI for Environmental Hazard Mitigation and Resilience: Pathways to a Safer Future*, explores how AI tools are being mobilized to forecast and manage environmental hazards, including floods and wildfires. Through predictive modeling and real-time monitoring systems, this chapter showcases case studies that illustrate the power of AI to improve risk assessment and adaptive response strategies. The ethical challenges of deploying AI at scale are also addressed, reinforcing the need for responsible innovation.

In **Chapter 7**, *Mapping Urban Planning and Sustainability: A Bibliometric Perspective*, the authors conduct a systematic bibliometric analysis of research trends in urban planning and sustainability. By reviewing nearly 900 scholarly documents using the Bibliometrix package, the chapter reveals key influencers, regional research patterns, and thematic evolution in this interdisciplinary field. This metanalytical approach provides a valuable overview of how sustainability and geospatial studies intersect in academic discourse.

Chapter 8, *Health and Habitat: Rights of Sundarbans Tribals in Flood Risk Context*, bridges environmental mapping with public health by focusing on the tribal populations of the Sundarbans. Using geostatistical techniques, the chapter highlights the vulnerabilities faced by these communities due to recurring floods and changing wetland landscapes. It calls for integrated policy frameworks that link ecological sustainability with healthcare access, ensuring that indigenous communities are not left behind in disaster response or climate adaptation efforts.

Together, these chapters converge to form a comprehensive dialogue on the role of geostatistical tools, artificial intelligence, legal frameworks, and social equity in mapping and managing flood hazards and wetland dynamics. Each contribution adds a unique perspective, reinforcing the need for interdisciplinary collaboration in facing one of the most pressing challenges of our time: environmental resilience.

CONCLUSION

As editors of this volume, we are proud to present a multidisciplinary collection of chapters that explore the intricate interconnections between technological innovation, environmental resilience, sustainable development, and social justice. Each contribution reflects a critical effort to address contemporary global challenges—ranging from flood hazards in Sub-Saharan Africa to the nuanced rights of farmers and indigenous communities in the face of ecological and legal transformations. Through the integration of artificial intelligence, machine learning, geospatial modeling, legal frameworks, and behavioral theories, the chapters in this book offer insightful pathways for designing resilient societies and ecosystems.

The breadth of perspectives in this volume highlights the urgent need for collaborative, data-driven, and community-centered solutions to mitigate environmental risks, foster equitable access to resources, and sustain future generations. From mapping vulnerabilities and informing public policy to advancing smart urban planning and health equity in climate-sensitive regions, the research compiled herein stands as a testament to the power of cross-sectoral knowledge exchange. As we navigate an era of increasing complexity and rapid change, we hope this book serves as both a reference and a call to action for scholars, policymakers, practitioners, and all those committed to shaping a more just, adaptive, and sustainable world. Also, This work was supported and founded by the Deanship of Scientific Research, Vice Presidency for Graduate Studies and Scientific Research, King Faisal University, Saudi Arabia [Grant No. KFU 251760].

Walid Chouari
College of Arts, King Faisal University, Saudi Arabia

Abdelkader Mohamed Sghaier Derbali
Taibah University, Saudi Arabia

Chapter 1
Mapping Flood Hazards in Sub-Saharan African Region:
Hindering Regional Sustainable Development

Wasswa Shafik
https://orcid.org/0000-0002-9320-3186
Dig Connectivity Research Laboratory (DCRLab), Uganda

ABSTRACT

This study presents an analysis of flood hazard mapping in the Sub-Saharan African region and its impact on sustainable development. Flooding remains a significant challenge in the region, exacerbated by climate change, rapid urbanization, and inadequate infrastructure. Effective flood hazard mapping, using geographic information systems (GIS), remote sensing, and machine learning, is crucial for disaster preparedness, risk mitigation, and urban planning. However, limited access to accurate data, insufficient technical expertise, and funding constraints hinder the development of comprehensive flood mapping systems. The resulting vulnerabilities lead to displacement, economic losses, and disruptions in agriculture, directly affecting food security and livelihoods. Addressing these challenges requires a multi-stakeholder approach, integrating policy reforms, community engagement, and technological advancements. This study explores the role of flood hazard mapping in sustainable development and climate-resilient infrastructure to mitigate flood risks and enhance regional sustainability.

DOI: 10.4018/979-8-3373-3206-2.ch001

1. INTRODUCTION

Floods continue to wreak havoc in many countries across the globe. In the wake of many natural disasters and anthropogenic activities that have triggered devastating floods, the work is engineered to evaluate the flood hazard across Sub-Saharan Africa using GIS multi-criteria decision-making techniques and satellite data from 2006 to 2015. The novel aspect of the current research is the creation of some unreported parameters like the basin lag time and the examination of 245 basins. Degradation in vegetation cover, road networks, and slope contribute significantly to the basins' vulnerability (Gudoshava et al., 2022). There is subsequently a weakening in the flood hazard as the depth, basin area, and discharges responsible for generating a flood hazard are ineffective in creating a warning system. The severity of the flood translates to the dams with the capacity to prevent catastrophic flooding by releasing water, contributing significantly to the propagation of the flood hazard. Though the analysis offers significant potential for improving the flood hazard across the SSA, assessments of the flood risk would require data on asset development and the economy (Leauthaud et al., 2013).

Considering the latest global changes, the Sub-Saharan African (SSA) region faces new socio-economic challenges like the rest of the world: an increase in population, development of new infrastructures, implementation of the digital economy, changes in lifestyles, and modernization of agriculture. SSA is the area and regions of the continent of Africa that lie south of the Sahara. These include Central Africa, East Africa, Southern Africa, and West Africa (Shafik, 2024b). In parallel, the region is characterized by a high vulnerability to threats. Several events can occur: rapid onset disasters, slow-onset disasters, natural disasters, and disruptions. In terms of socio-economic damage, the trend is increasing over time. Risk-related disasters threaten the ability of societies to achieve their objectives. They can impact the population physically and/or materially, as well as economic, social, cultural, and environmental issues (Shafik & Kalinaki, 2024).

Among natural disasters, floods are the most frequent of all natural hazards. In West and Central Africa, for example, the number of floods has increased, and their impacts are straining development efforts. A number of hazards, such as rapid urbanization and deforestation, combined with a lack of adequate infrastructure and emergency services, are making these floods more dangerous and devastating to the population. Since the last decades, environmental issues have been at the heart of concerns to ensure a sustainable future for humanity (Chilunjika & Gumede, 2021). Preserving natural resources is a priority, but the protection of populations and breeding areas is equally important. The proposed research work aims to use hydrological modeling in combination with remote sensing data to analyze flood hazards over the Sub-Saharan African region. Information products represented by hydrological indicators should be used to strengthen the decision-making process for the resilience and adaptation of populations. Additionally, the knowledge generated by our multidisciplinary research effort will provide insights that can help prevent the material and human losses that SSA (region illustrated in Figure 1) experiences as consequences of flood events (Gudoshava et al., 2022).

Figure 1. Sub-Saharan Africa includes West Africa, Central Africa, East Africa, and Southern Africa (guesthollow)

1.1. Research Objectives

These research aims are discussed in the following order. The chapter gives a brief discussion of the aims of the study. An extensive literature review follows this. The literature review includes a discussion on why it is important to explore the rainstorms, floods, and landslides in the Sub-Saharan African Region. The scenario development, expert consultation, and multi-criteria decision analysis to evaluate the risk hazards are then discussed. Elucidates the characteristics of the Sub-Saharan African Region. We briefly introduce the different biomes and climates encountered in the South African region. The different types of potential and causes of disasters are then discussed. At the end of the discussion, we pinpoint the importance of the research. For ease of reading and understanding, a conclusion summarizing all the studies is presented. The principal aim of the study is to develop a set of qualitative and quantitative scenarios to think through the issues of rainstorms, floods, and landslides in the Sub-Saharan African Region. The objectives are to (1) develop an integrated set of qualitative and quantitative scenarios that are plausible and relevant to the different stakeholders that would face rainstorms, floods, and landslides in the Sub-Saharan African Region and (2) determine the magnitude and likelihood of these hazards.

1.2. Scope and Significance

Sub-Saharan Africa now faces many development challenges due to population growth. It is the size of 8,000 craft lands without permanent access for 700 million inhabitants and 60% of the world's underdeveloped countries. Water-related disasters such as floods are significant obstacles. Heavy rain is the most frequent natural disaster in the sub-region, which results in flooding. Flood disasters have become increasingly frequent in recent years, causing widespread loss of life and property while undermining many development activities (Leauthaud et al., 2013). Greater hazards are possible if flood hazards are combined with poverty, population growth, and land development issues in Sub-Saharan Africa. Despite alarming predictions of a significant increase in the number of floods in sub-regions and worldwide in the future, there is little effort to develop and integrate a reliable hazard monitoring and management system (Leauthaud et al., 2013). This is our main motivation for establishing a flood hazard system covering various geographic areas within certain longitudes. The attention paid to the complexity and dynamics of the ranked sub-regions of the earth and the support from the broad technical and economic community also inspired our team to focus on this research.

2. FLOOD HAZARDS IN SUB-SAHARAN AFRICA

The continent of Africa is vulnerable to several natural disasters, including floods, particularly in the sub-Saharan region. Flood hazards have increased over the years, and a disproportionate number of such disasters have caused havoc, ravaging livelihoods and economies in the region. Flood damage and dangers can increase, thereby affecting more people, with few benefiting from upgrades in the standard of living and reduction of fatalities or threats. Essentially, the number of deaths from floods has reduced due to improved preparedness; however, there is a lack of access to facilities providing clean drinking water and security, while recovery from economic loss for those affected is insufficient (Ibrahim & Mensah, 2022). Floods are the harshest natural disaster that affects more people than any other type of natural disaster. It was also observed that sub-Saharan Africa frequently encountered larger flood events in the four years from 2000 to 2003, which cost approximately sufficient to develop social and infrastructure needs. The flood risk is expected to increase by 12–80% by 2050, depending on the area subjected to flooding. The capability of sub-Saharan African countries to manage floods is constrained due to the region's vulnerability to climate change hazards. Most of these hazards intersect, whether it is a routine flood event or a devastating one (Simatele et al., 2012). With growing competition for water and agricultural areas, upgrading universities and sub-Saharan Africa's cities are experiencing stronger rainfall, which causes the surface to change into a rising level of water in the rivers, increasing the risk during flood events.

2.1. Causes and Impacts

Floods are the most high-impact natural disaster in the entire range of human disasters. These events are spatially and temporally random in occurrence, while they are caused by either slow-onset processes or unusual meteorological conditions. Although no particular classification of floods can suit every purpose, floods have been naturally stratified by a combination of their original causes, hydrodynamic manifestations, and governing environmental high-impact implications: impacts upon flood-depth probability estimates, depth-damage functions, floodplain zoning, pollution generation, response times, warning adequacy, adaptation planning, and criteria for schemes' long-term sustainable survivability, among others of less immediate practical importance (Dossou-Yovo et al., 2023). The most common causes of floods in the Sub-Saharan African region are rainstorms, hurricanes, cyclones, or tropical storms, steep topography, urbanization, snowmelt, agricultural practices, waste-blocked drainage, wind-driven tides, encroachment within floodplains or coastal zones, slow glacier plugging discharge, overtopping of flood-control structures, and climate change or geohazards, among others. The direct impacts, indirect consequences, lag effects, and results of the floods, particularly damage to building infrastructure of economic worth, also need to take account of topographical variation within the urbanized area in question (Perez-Saez et al., 2022). Although flooding is a natural environmental phenomenon, planning, design, construction, and development activities generally have an impact on the environment and the downstream catchment area, and this, in turn, will have implications on the scale-up of the larger infrastructure-related scheme.

2.2. Vulnerability and Exposure

The need to maintain and improve the livelihoods of the population in the African region forces most of the population to develop their living spaces in high-risk sites. These high-risk zones have often never been studied. The overflow of rivers, in particular the lack of regular maintenance of riverbeds and the deforestation of watersheds, accentuates the flood hazards, which add the risks of drying riverbeds, erosion, tailings from quarries and dams, and pollution of groundwater (Echendu, 2021). The consequence of these hazards is not only to slow development but also to provoke irreversible degradation of the environment and increase poverty. The point raised is to search for harmonious development between the population, the environment, and economic development in zones exposed to hazards. It is up to the environment and development study group researchers to anticipate and propose various solutions, taking into account the various players in the economic development chain (Jongman et al., 2012). The African region is continually undergoing many hazards to overcome. In fact, because of their rapidly increasing population, the problems often come from man himself. In the African region, the behavior of the population in very high-density areas, such as the deforestation of watersheds, the landscape, and the basic materials, along with the lack of use of equipment for various needs, confuses the actions of environmental research and development in these zones with the hazards they cause. The hazardous zone becomes a phenomenon, and the population spends a significant part of the budget on repairs, which adds a risk of aggravating poverty (Djègbè et al., 2020). This result impacts intellectual, cultural, spiritual, and moral life. The living environments, the streets, and the sordid landscape are transformed into places of degradation in which the population lives in a promiscuous way with garbage. These spaces sometimes reveal the exclusion and segregation of certain parts of the human population. Finally, vulnerability materialized by diseases, accidents, and epidemics (Li et al., 2021).

3. EXISTING MAPPING TECHNIQUES

3.1. Traditional Early Warning Systems

Floods in the Sub-Saharan African region are often treated primarily as a disaster issue to be managed after an event has occurred. Many regions, including Ghana and Nigeria, are using traditional early warning systems that rely heavily upon agent-delivered forecasts supplied by traditional weather services and local environmental chambers. These forecasts are then verbally communicated to end-users, and prescriptive actions are developed and implemented. Similar systems are in place in many resource-poor environments (Møller-Jensen et al., 2023). The systems utilize local experiences or studies of historical weather conditions, such as river water levels or flow, water or mud levels, soil moisture, or river velocities, to instruct action. While these alert systems have saved lives, destruction to property remains high. Simplistic or poorly grounded warnings lead people to ignore the messages or believe that they have effectively prepared, putting them at risk of being killed or suffering major injuries from the flood (Anami et al., 2009).

3.2. Flood Modeling

The adoption of numerical models for improved real-time weather and climate forecasts and increasing computing power has increased awareness of using hydrodynamic models for real-time or near real-time river flow forecasting; however, the process is lengthy and requires data that are often not readily available outside developed regions. Developing and calibrating such solutions can be time-consuming, and therefore, academic hydrologists often avoid the subject. The prominence of and interest in flood models during recent years have increased their use, with many hydrodynamic model technologies for developing flood maps based upon slow, accurate studies that support infrastructure design and planning decisions (Ibrahim & Mensah, 2022; Leauthaud et al., 2013; Li et al., 2021). An emerging area of research is to translate these maps into real-time maps to encourage their use by the public. The process of using rainfall-runoff models to enable discharge measurements to be turned into flood histories that can be amalgamated into historical maps of frequency and impact is being explored. Canonical examples of this type of work outside the United States are in West Africa (Li et al., 2021).

3.3. Risk-Proximate Mapping

The last common form of flood mapping attempts to directly map the flood risk. This type of information is extremely valuable for economic development and disaster planning applications. Corporate economic planners and investors use such information for site selection and to negotiate insurance costs, while emergency response planners use the information to develop evacuation and resource placement strategies and to assess threats to public health and safety. Most regions lack such maps, and African regions have the least number (Geris et al., 2022). Flood hazard maps blend the output of the rainfall-runoff models and a consideration of likely changes in land use and future climate to develop discrete mappings indicating what parts of a drainage basin are at greatest risk of being flooded. Just as with historical and modeled maps, the level of detail of information is a function of the map's purpose and scale. Larger-scale maps typically indicate where high watermarks are most likely to be found, which structures are most likely to be inundated, and the likely effects on infrastructure; smaller-scale maps might display a single iso-line representing 100-year flow (Bukari & Aluko, 2023). Flood hazard maps complement modeled flood history maps and flood frequency maps, each of which describes different parts of the complex flood hazard.

3.4. Remote Sensing and GIS Technologies

Remote sensing and GIS technologies have evolved to become primary instruments for environmental research and investigation. Technology continues to present a significant and powerful force in combating those problems that hinder sustainable development. The cloudless visibility that technology provides has contributed to the increased production of digital data, thus supplying the vast data needed for sensor-based applications and providing an ideal opportunity to monitor, understand, and address environmental problems (De Perez et al., 2017). Sensing-derived environmental datasets form the backbone of GIS, permitting the portrayal and examination of spatial and attribute data that model an incredibly precise representation of the physical world. Versatile and powerful, GIS offers a comprehensive and holistic approach to solving environmental issues, which include various geomorphological and ecological issues related to flooding. The rapid deployment of sensing technology, which offers real-time data processing, is significantly beneficial (Shah et al., 2022). For flood management and control, remote sensing and GIS technologies are deployed in monitoring and evaluating flood incidents through the formulation of proper relief plans, assessing hazards, formulating risk assessments, and advising on the preparedness and response steps to be taken in the event of flood hazards manifesting in hazardous consequences (Serdeczny et al., 2017). Other essential applications of the technology include modeling surface runoff and the quantification of extreme events, which include forecasting and providing available observations. Remote sensing technology processes include data analysis and visualization of data captured from the study area. The outputs include thematic hazard maps, analyses of change that lead to the recognition of areas that are more susceptible to flooding, and the establishment of planning guidelines to manage flood hazards (Müller et al., 2023).

The major advantages of specific operations and the results of the flood vulnerability and risk assessment are rather obvious and include the generation and verification of bathymetric maps, the determination of potential areas and infrastructure exposed to flooding due to severe hydrological events, and possible sources of pollution or dangerous contaminants released into the environment. Although the assessment results are not exact due to inherent constraints in models and data input, the results present planners and policymakers with a valuable alternative for choosing from a number of flood protective measures. Indeed, reliable flood data is essential for properly constructed decision-making and adequately designed sustainable development (Uttajug et al., 2023). Limitations representative of this hydrological module working within a GIS environment are mainly in the smaller amount of existing and good-quality data in the region, especially in the domain of river conditions, sediments, and seasonal variations, the lack of hydrological stations necessary for the in-situ calibration of transforming functions and real-time flood assistance, silting of major river beds, and shrinking of smaller rivers due to climate changes, and the excessive presence of high dikes surrounding inhabited areas, promoting preventable flooding of immediately adjacent areas. These technical and data deficiencies are aggravated by the existing water management model based mainly on the knowledge and experience of local experts. Data from satellite monitoring is not common and, when available, is hardly accessible and affordable. Data existing at the regional and state authorities are vague, often out-of-date, and fragmented, while unified basin monitoring and control is still at the level of utopia (Rentschler et al., 2022). The lack of complicated software and related specialists is another obstacle in the practical exploitation of these findings, and significant training is needed, as presented in Table 1.

Table 1. Key factors contributing to flood hazards in Sub-Saharan Africa

Factor	Description	Impact on Flood Risk	Affected Regions	Potential Solutions	Reference
Climate Change	Increased rainfall variability and extreme weather events	More frequent and severe floods	West and East Africa	Climate adaptation policies, mitigation strategies	(Gudoshava et al., 2022)
Deforestation	Loss of forests reduces natural water absorption	Increased surface runoff and erosion	Central and Southern Africa	Reforestation, afforestation initiatives	(Leauthaud et al., 2013)
Urbanization	Unplanned settlements with poor drainage systems	Waterlogging, flash floods	Lagos, Nairobi, Kinshasa	Urban planning, improved drainage systems	(Ibrahim & Mensah, 2022)
Poor Land Use Planning	Settlements in flood-prone areas	Increased exposure to flood hazards	Riverine and coastal regions	Zoning regulations, floodplain management	(Simatele et al., 2012)
Lack of Early Warning Systems	Limited technology for predicting and warning communities	Delayed response, higher casualties	Rural and informal settlements	Investment in forecasting tools, community awareness	(Dossou-Yovo et al., 2023)
Inadequate Infrastructure	Weak flood control measures (dams, levees, drainage)	Poor water management, infrastructure failures	Low-income urban areas	Investment in resilient infrastructure	(Perez-Saez et al., 2022)
Soil Degradation	Erosion reduces the soil's ability to absorb water	Faster runoff, worsening flood impacts	Sahel region, Ethiopia	Soil conservation techniques, agroforestry	(Echendu, 2021)
Rising Sea Levels	Coastal erosion and saltwater intrusion due to climate change	Increased coastal flooding	West African coastlines	Coastal defenses, mangrove restoration	(Jongman et al., 2012)
River Overflow	Excessive rainfall causes rivers to exceed capacity	Flooding of agricultural and residential areas	Nile, Congo, Niger basins	Riverbank protection, controlled water release	(Anami et al., 2009)
Limited Disaster Response	Weak institutional capacity to manage floods	Slow recovery, long-term socio-economic losses	Various national governments	Strengthening disaster management policies	(Tiepolo & Galligari, 2021)

4. CASE STUDIES

This section outlines four case studies of the application of the algorithm termed Bakar modified K at present for mapping flood hazards at a global scale. This approach is a worldwide resolution and computationally efficient for obtaining instantaneous flood spatial distributions due to the inclusion of existing observations.

4.1. The Indus River, Pakistan

We present the application of the proposed algorithm for mapping the Indus flood hazards. On April 5, 2014, a supervised classifier model was constructed using the results of Bakar modified K. The input data for training purposes were obtained from April 5, with data and corresponding soil moisture realizations acquired from the satellite dataset. The land cover and land use training samples were created as a mask by the unsupervised algorithm in the vicinity of the analyzed area. These samples were used to train the developed unsupervised land classification method (Müller et al., 2014). In accomplishing a model for land cover and land use classification, an extended area covering around 20, 78, 30, and 75 was considered for the testing stage. The characteristics required for fitting the classification approach were finally produced.

4.2. The Amur River, Russia

The next application is on the highest Amur River on September 1, 2025. This domain was modified considering the width of the Amur River (the area taken into account in this case study equals the hydrographic basin area of 540,000 km^2) and includes a region related to a second-level spatial resolution of the future requirements based on a maximum of 300 floods, 20% error, with an 8 km by 11.17 km resolution of inundated areas. Data resembling the future mission were obtained through the courtesy of a university (Dossou-Yovo et al., 2022). A priori, the depth bandwidth is fixed at 1 m (30 classified categories are considered). Cost parameters, indicating the relative importance of classifying each foam by identifications of the confusion matrix under consideration, are identified as values that slowly increase the balanced error of classification. In this study, we considered value 10 (high importance of foam classification, medium importance in class identification). The irregular shape is always selected using an empirical approach (Rother et al., 2022). The final product is displayed and evaluated by comparing it to observations.

4.3. SSA Country-Specific Approaches

Adapting the global data-driven model to specific countries offers the opportunity to capture the unique hazard profile, poverty dynamics, and other socio-economic, demographic, and geographic characteristics of a country. For instance, models at smaller spatial scales, lower computational complexities, and that consider data availability constraints can be developed for specific countries. The risk model must capture relevant data for flooding risks and the relationship between the factors and the risks. Given the difficulty of collecting such granular data across Sub-Saharan Africa due to the prevalence of informal settlements, remote sensing and geoinformatics could offer essential tools for gathering the data (Ramiaramanana & Teller, 2021). Next, dynamic behavior can be endogenously included in the model to account empirically for the interactions among the different factors influencing the risk of flooding. Such dynamic models could capture the relative advantages of low-income countries in response to specific external events that induce flooding. For example, structural positions and urban form influence risk perception and can lead to undesirable outcomes of increased flood risk in the face of increasing exposure to housing developments and water demand driven by unfilled urban jobs. Third, the risk model should highlight possible transmission channels and critical factors influencing the process through a well-defined theoretical model (Njoku et al., 2015). Fourth, structural factors, behavioral factors, extreme weather events, and other factors elicited directly from local data and additional expert interviews can be incorporated into the model-building process.

In order to map flood hazards in the Sub-Saharan Africa context, data availability, together with potential modeling choices, might limit the likelihood of obtaining a GIS hazard dataset. Sub-Saharan Africa represents the region with the least investment and capacity in flood risk analysis. African flood models in urban contexts are few and represent mostly heuristic methodological approaches. Furthermore, multiple proprietary global and pan-African geospatial datasets exist, including normalized difference vegetation index and other derived datasets (Nagata et al., 2022). Choose freely available data sources and leverage previous international or academic research to comply with open scientific research principles. These principles can also be applied to remote sensing data, maps, digital elevation models, climate model outputs, and other GIS-related inputs that are commonly provided by public and private entities. No-cost trade-offs come with the increased flexibility of the open-source datasets, which drastically lower data acquisition overhead typically associated with natural hazards databases, necessitating a cost-weighted approach to the selection of datasets that best-fit research objectives (Gannon et al., 2018). Introduce to the hazard model context additional local information from field observations, household survey data, community focus group discussions, and expert interviews to achieve a higher context awareness. Disproportionate modeling research funding often leads to disproportionate research results. Remote sensing data is a highly effective GIS tool for identifying potential indicators of harm, exposure, vulnerability, and adaptive capacity before and after the fact in the modeled context. The wide variety of available, free, user-friendly remote sensing processing software tools allows for the establishment of sophisticated workflows that are easily controlled. The quality of remote sensing data will refine the planning of field expeditions and reduce costs while ensuring that the field team can be in the field for as short a time as possible, allowing the teams to conserve more energy for field data collection and record-keeping upon arrival at the study site (Codjoe & Atiglo, 2020). When fitted to the modeling context, data sources can be used as a cornerstone for data analysis and scientific reporting.

4.4. Success Stories and Challenges

Most of the large-scale flood hazard and risk mapping in this region has been carried out in South Africa. Success stories, therefore, are drawn mostly from the South African experience. Available studies regarding the use of remotely sensed data in floodplain mapping and risk assessment exist for the Berg and Liesbeek River catchments. The methodologies used in these studies are different, but there are clear conclusions that aerial photography can provide an important source of information that increases the accuracy and reliability of floodplain delineation (Akinsemolu et al., 2024). These kinds of studies regarding different urban areas are more often missing for the region. A relative scarcity of aerial photographs may trigger this, but more properly, knowledge transfer or lack of understanding of the capabilities of either remote sensing images or other available technologies to capture required datasets. Yet at least two additional studies from South Africa that use remote sensing, and geospatial technologies have dealt with the Kaalspruit flood. These comparative studies between the manual approach and the numerical model-derived model proved greater effectiveness and speed of determining the spatial extent of the flood and established benefits underscore the effectiveness of the use of GIS-based technologies in flood management tasks (Neumann et al., 2015). Whilst these case study examples, where there is a strong link of remote sensing technology with on-the-ground empirical flood delineation, are found for South Africa only, this summary clearly identifies a significant opportunity in the use of such technologies for risk assessment and mapping of the flood hazard, as illustrated in Table 2.

Table 2. Strategies for sustainable flood hazard mapping and mitigation

Strategy	Description	Technologies Used	Expected Outcome	Implementation Challenges	Reference
Remote Sensing & GIS	Satellite imagery to map flood-prone areas	GIS, drones, satellite data	Accurate flood risk assessment	Data accessibility, cost of implementation	(Negev et al., 2019)
Hydrological Modeling	Simulation of water flow and flood impact	SWAT, HEC-RAS, MIKE Flood software	Prediction of flood-prone zones	Model accuracy, need for real-time data	(Mubangizi et al., 2023)
Community-Based Monitoring	Engaging local communities in hazard identification	Citizen science, mobile apps, social media	Increased awareness, faster response	Lack of community engagement, literacy barriers	(Lemaire, 2021)
Early Warning Systems (EWS)	Real-time flood alerts and prediction	IoT sensors, AI, SMS alerts	Reduced loss of life and property	Infrastructure costs, network coverage	(Joshua et al., 2022)
Sustainable Land Use Planning	Regulating development in high-risk areas	Urban planning tools, policy frameworks	Minimized flood impact on communities	Resistance from developers, enforcement issues	(Cuthbert et al., 2019)
Green Infrastructure	Nature-based solutions like wetlands restoration	Mangrove reforestation, permeable surfaces	Improved water absorption, reduced runoff	High initial costs, long-term maintenance	(Griffith et al., 2006)
Flood-Control Infrastructure	Construction of levees, dams, and flood channels	Engineering solutions, smart water systems	Controlled water flow, reduced flooding	Environmental impact, funding constraints	(Uttajug et al., 2023)
Climate Adaptation Policies	Government strategies for resilience-building	Policy frameworks, economic incentives	Strengthened regional disaster preparedness	Political commitment, coordination issues	(Müller et al., 2023)
Data-Driven Decision Making	Using big data and AI for flood risk analysis	Machine learning, geospatial analytics	Better flood prediction and resource allocation	Data privacy, integration with local policies	(Shah et al., 2022)
Public Awareness Campaigns	Educating communities on flood risks and preparedness	Radio, TV, social media, community workshops	Empowered communities, proactive response	Reaching remote populations, misinformation	(Shafik, 2024c)

5. POLICY AND GOVERNANCE

Sub-Saharan African countries have seen an increase in the frequency and associated costs of flood disaster events. This flood risk is due to a combination of several triggering factors, including but not limited to environmental, social, and climatic issues. The region's development efforts increase the population's exposure and vulnerability; flood risk mitigation can have an undesirable impact on the development process, especially if it is not well managed and integrated into the regional and local development plans (Shafik, 2024a). Understanding the flood-defined hazard baseline and its possible impact in a given area, including how it could affect people, is essential for countries' disaster risk management. However, this understanding is hampered in Sub-Saharan African countries by a lack of harmonized and comprehensive methods that are tailored specifically for the network of local knowledgeable factors affecting exposure and vulnerability dimensions of the flood hazard, with implications for regional and local effective governance, preparedness, and response (Shafik, 2024d).

Unqualified and non-flood-specific hazard evaluation is majorly responsible for the high vulnerability rate that has been attributed to flood disaster events in this region, where the absence of such a comprehensive method and/or its non-integrated use compromises the effectiveness of public policy and private individual or collective decisions devoted to different local uses of flood hazard-related resources throughout the different regional and local levels of government. Moreover, the questions of why, how, who, and what coping and land use strategies remain largely unanswered when addressing the flood hazard landscape in particular, enhancing the development-disaster risk spiral in general (Codjoe & Atiglo, 2020). Furthermore, rapid development throughout the Sub-Saharan African region dictates that the hazard should be routinely reassessed. In a perfect world, both the first-time definition and periodic reassessment should not only incorporate an increased intrinsic scientific understanding and structural improvements but also reveal any methodological inconsistencies, methods that do not adapt to the users' community, and semantic, language, and communication problems which often crosscut the entire hazard assessment process. Instability in the acquired knowledge base casts uncertainty on the overall quality of development decision outcomes (Leauthaud et al., 2013). Consequently, the flood disaster risk level – a multiplier of the quantitative measure of the hazard and the number of expected disaster occurrences – increases.

5.1. National and Regional Strategies

At the national level, the following specific efforts may ensure the incorporation of geospatial tools and systems related to flood risk preparedness, forecasting, and mitigation, fostering sustainable development efforts in the Sub-Saharan African region. In developing countries, where the government structure can facilitate collaboration, it is especially important to comprehensively engage the public, private, and academic sectors to implement national geospatial tool use. Develop national, regional, sub-regional, and local geospatial maps wherein hydrological modeling and analysis, and flood hazard maps prioritize the key flood-prone areas to assist decision-makers in developing improved land management plans (Bukari & Aluko, 2023). Organize integrated flood risk management initiatives controlled by governments in countries like SSA. Support sharing and access to necessary donor institutional facilities to reduce transaction costs and promote the integration of geospatial data, tools, and systems in subsequent technology transfer and cooperation initiatives vital for development that promotes partnerships between developing countries. Consider including partnerships with other providers as part of the strategy to harness indigenous capabilities. Promote broad general support from local action leaders and tailor methodologies to prepare, evaluate, and adopt geospatial tools for mapping and assessment of flood hazards. Therefore, institutions provide an ideal entry point with dedicated mechanisms for technology awareness and dissemination, technology needs assessments, and interactions (Müller et al., 2023). The sub-region economies, on the other hand, have the responsibility to budget for and set up entities at the national, regional, sub-regional, and local levels, envisaged to provide the financial resources and strong local presence required to offer direction, inspire, and build technological capacity, provide incentives for domestic technology, as well as strengthen human capital essential for the promotion of flood risk preparedness, forecasting, and mitigation enabled by geospatial tools and systems crucial for sustainable development. In order to develop a conceptual framework for attaining optimal development structures, a fingerprint of the mini framework for sustainable technology development is presented in this section (Shah et al., 2022). This fingerprint provides a common basis for analyzing the performance of development structures in promoting sustainable development.

5.2. International Collaboration

Acknowledging that the capacity of low-income countries and lower-middle-income countries is developing and needs support to develop national and regional flood infrastructure, tools simulating flood impacts, and knowledge about data availability and readiness for use. There is also a significant international focus on disaster risk reduction but limited financial and human resource support curtails effectiveness. However, the impacts of hydro-meteorological hazards at the local level are interconnected with spatially detailed rainfall, land cover, topography, and weather parameters at finer scales, knowledge, and expertise that are primarily held at high-income academic institutions and private industry (Mangani et al., 2018). In this context, such "big data" can be used to support development in Africa under global collaboration. Investing in Africa's water infrastructure is essential for meeting the targets of the United Nations SDG Goal 6 and thereby supporting progress towards all the SDGs. Building on existing international collaboration and funding platforms, international investment will help accelerate the needed capacity convergence that will develop and sustain the necessary infrastructure that will reduce African flood vulnerability. Combining international collaboration with improved data sharing can lead to increased data quality and several benefits relating to socioeconomic development, policy, governance, and the strengthening of fragile states (Li et al., 2021). Strategies must also aim to address the gender gap and promote inclusive, sustainable development, especially in vulnerable regions that are particularly susceptible to adversities like flooding.

6. CHALLENGES AND OPPORTUNITIES

Mapping of hydrological hazards, particularly in the Sub-Saharan Africa region, is limited due to data gaps. The availability of hydrological data is key to the development and operationalization of hydrological hazard risk models. The lack of accurate hazard maps hinders precision decision-making by policymakers and society in general. The development of novel approaches based on rainfall and topography data can provide information on the occurrence of flood events in low data coverage regions (Anami et al., 2009). Anticipating flood-prone areas could then lead governments, international organizations, and NGOs to effective planning of policies on land use, agriculture, and prevention of flooding such as diking the flood banks, establishing reservoirs, as well as pre-positioning humanitarian supplies. In many flood-prone areas, critical pieces of information are difficult to obtain. Many of these areas are located in low- or middle-income regions with little funding for continuous improvements in hydrological monitoring through traditional physical observation methods. Furthermore, a combination of socio-political and infrastructural factors is coupled with low population density and governance failure in these less developed areas, making it difficult to apply the traditional physical and economic frameworks and justify investment in infrastructure to abate natural hazards (Tiepolo & Galligari, 2021).

6.1. Data Gaps and Inconsistencies

The RS-GIS technique requires huge volumes of data collected from various sources for geophysical, geomorphological, environmental, and socio-economic evaluation and analysis. Information gathered from field mapping, regional scale, and catchment scale studies, as well as remotely sensed and GIS data, lacks the necessary uniformity, level of precision, and temporal resolution required for integration into flood potential analysis (Müller et al., 2014). This will result in difficulty in setting up the modeling flow for cadastral subdivisions, errors in linking the globe to the local-scale observations, and in calibrating the models. This is due to data insufficiency, especially for developing countries and remote and topographically complex regions with a high degree of rainfall-induced floods. Land use and land cover, the multi-date and multi-sensor imagery, gauged streamflow or flow accumulation, stream channel width, and other physical parameters are required for calibrating and validating the model. Inadequacies of available and past information may distort the model structure, e.g., where urbanization encroachment and climate change effects are essential but not documented (Rother et al., 2022). Additionally, the situation is made more complex by the necessary connection to the international standards for a global vision that does not exist yet.

6.2. Capacity Building and Technology Transfer

International support helps in addressing many of the human capacity development challenges in Africa. For example, in April 1998, a global program on flood management was established, aiming to help Sub-Saharan African nations manage the risks associated with floods and address some of the key factors impeding sustainable economic development in the region. The initiative has developed capacity building materials to help focal points better coordinate and prioritize assessment needs and to help develop flood mitigation strategies tailored to meet the specific requirements of each country. Through these training programs, participants can become familiar with flood management techniques and the use of the information and guidelines provided (Adelekan, 2020). Training is also seen as a means of facilitating transfer and promoting the use of appropriate technologies, approaches, and best practices for sustainable development. The need to transfer environmentally sound technologies dominated discussions in 1992, and principles on technology transfer are referenced in a number of international treaties. The transfer of technology to developing countries is a recognized tool to address social and economic development in flood-prone and affected areas. While there has been an increase in the number of initiatives and mechanisms for technology transfer through training, knowledge, and experience, direct funding has also been provided to the national disaster management centers to build capacity. A special fund has been established to support the improvement of systems for collecting disaster-related data, preparedness for natural disasters, and response to natural disasters (Ramiaramanana & Teller, 2021). The support of the Regional Centre for Mapping Resources for Development in the modernization of member states' land administration systems is part of these rural development measures.

7. INTEGRATION WITH SUSTAINABLE DEVELOPMENT GOALS

The process of mapping flood hazards in the sub-Saharan African region will no doubt be a milestone toward several sustainable development goals, not limited to the Amajuba district and eThekwini, but also the national and global agenda on sustainable development. Mapping land characteristics and determining areas susceptible to flood hazards can be overlaid with poverty or unemployment attributes. This information would feed development agencies, local authorities, and regional planners with valuable pointers for adaptation and planning purposes, ratepayer allocations, and reconstruction and development programs. Employment opportunities can thus be found throughout the cycle of development through the construction, monitoring, and maintenance of flood defense systems, enhancement of increasingly valuable wetland areas, and efforts to secure adequate housing for some of the poorest inhabitants of districts from the destructive influences of flooding (Njoku et al., 2015). Sharing resources reduces risks and costs, is likely to stimulate local buy-in, and will promote a just and balanced criteria for decision-making about where and how people live, and their right to protection from predictable, frequent, and severe meteorological extremes. It is not the implementation but the timely dissemination to those in need that remains the primary stumbling block for successful early warning and disaster prevention strategies. The future development of the sub-Saharan African region and its urban and rural areas will be primarily predicated upon both its and global ability to contain the rising burden and complexity of its current disasters, which constitute a drain upon resources. These events must be curtailed, or the ability to proffer assistance and development assistance will be progressively weakened (Gannon et al., 2018). The political, economic, scientific, and institutional strategies necessary to reverse these trends are numerous; the commonalities of interest are almost as conjugate, and only heightened awareness of the need to understand and mitigate these disasters will inform the scale and speed of the implementation of such transformation.

7.1. Linkages and Intersections

Paradoxes of poverty and richness in natural resources characterize Sub-Saharan Africa. It bears the brunt of natural extremes such as floods and droughts. Hence, even when the continent has made some gains in development goals, these have been constantly undermined by recurring natural hazards. The extent of flood mapping in the region has been reviewed to understand what is available, help establish baselines and assist policymakers in prioritizing evidence-gathering to drive decision-making in sustainable development (Codjoe & Atiglo, 2020). The available information has been grouped based on the nature of the flood hazard and described the varied sources providing respective metadata. Some of the available datasets have some factors disaggregated to enable linkages and intersections with other datasets and allow the scope to carry out research analyses. Covering just over 20% of the continental landmass, the Eastern and Southern Africa Water Science Network maps have been used quite extensively in the region. It appears that the database was richer in areas that had experienced relatively historical flood events in well-watered parts of the Sub-Saharan African region. Coverage of the relative surge flooding parameters presented in the physical vulnerability papers and the humanitarian atlas of flood parameters has helped traverse into other very useful analyses, such as socio-economic vulnerability (Neumann et al., 2015). Datasets that do not provide factors needed for intersections lack the scope to carry out research analyses, which are instead dependent on overlaying multiple maps for floodplains, population, rural and urban extents, land use, property, infrastructure, stakeholder interests and activities, culture, and social networks.

8. FUTURE DIRECTIONS

This study only dealt with flash flooding and considered rainfall as the only meteorological parameter responsible for the onset of flooding in the Nigerian region. To better adapt this methodology for mapping flood hazard intensities, it will be useful to incorporate remotely sensed hydrometeorological information such as sea surface temperatures, global atmospheric temperatures, heights above sea level, and atmospheric moisture, including atmospheric water vapor throughout the entire atmospheric boundary layer (Lemaire, 2021). Such inclusion can result in the identification of regions having high values of these parameters associated with a given flooding event, which may lead to the development of a relative flood hazard intensity index map. Additionally, including remotely sensed precipitation data from other satellite-based platforms in the proposed methodology can serve as a useful data source for accurately estimating the meteorological parameters that trigger flooding disasters across Nigeria and any other region of the globe (Griffith et al., 2006). The incorporation of remotely sensed SO2 and aerosol optical thickness derived from other sources into the methodology as anchoring geoinformation datasets can aid the detection of atmospheric convective initiation, which can broaden the application scope of this approach.

8.1. Innovative Technologies and Approaches

Innovative technologies and approaches such as citizen science or crowdsourcing for improving spatial data sharing can address some of the challenges of spatial data sharing. Crowdsourcing has facilitated spatial data sharing with dynamic and more real-time spatial data updating. The lack of good communication technologies or infrastructures is a common challenge of citizen science or crowdsourcing. Social media data, both the texts and the resulting sensory data, are contributing to the progress of flood hazard mapping initially for spatial data sharing. Researchers have realized the potential of social media for fighting floods and are developing algorithms to identify and map real-time or near real-time flooding information for accurate city- to country-level flood monitoring (Müller et al., 2023). Social media data use is providing temporal improvements in flood monitoring compared to existing resources. The temporal nature of social media content is contributing to more accurate flood monitoring, which no existing global or regional system can guarantee. Either existing global systems are limited to daily monitoring or are not free at the moment of necessity, which further complicates the temporal monitoring of flood cases that surpass 24 hours. Social media are reducing the temporal aspect of near real-time flood monitoring to minutes and not hours. Validated flooding data directly crowdsourced from citizens' actions during the flooding event are being combined with images from sensors present in the crowdsourced social media posts (Griffith et al., 2006; Müller et al., 2023). Flood event-collected data, after a validation step, are used with statistical algorithms in producing the flood maps.

8.2. Research Gaps and Priorities

Despite some significant progress, our review of flood studies in Africa reveals that there remain many underexplored areas. This literature review calls for more focus and research effort in studying flood vulnerability rather than only the physical or excessive precipitation-based debate. To achieve the Sustainable Development Goals in the Sub-Saharan African region, it is important to understand the links between flood vulnerability and response, especially in terms of its multiplicative effect with other associated heterogeneous vulnerabilities (Mangani et al., 2018). Little is known about interconnected climate impacts associated with rapid urbanization and socioeconomic change that drive the increase in flood risk, particularly in terms of the low adaptive capacity of the vulnerable urban poor. This is an essential area for further research since Sub-Saharan Africa is the fastest urbanizing region in the world and is expected to experience the largest urban growth rates during this century. There is a lack of sustained attention to connected yet distinct dynamics that generate flood issues, emphasizing the disconnect between different disciplines. There is also an absence of land cover prediction, human activity change, amenity, and deriving flood risks, patterns, and the timing of these events (Dossou-Yovo et al., 2022). Accelerating urbanization may result in a significant increase in social and institutional inequalities, and research is needed to understand the social exclusion impacts resulting from the interaction between these and flood events (Akinsemolu et al., 2024).

9. CONCLUSION

The question of sustainable development is important, and strategically, capturing and modeling flood risk in an innovative and improved manner is necessary to safeguard the impact of anticipated investment and support the achievement of sustainable development goals so dear to the people of sub-Saharan Africa. The results and observations in this study are similar to those of many other studies conducted in both developed and developing countries. While the dynamics of flood hazard might differ from the developed world, understanding the complexity of these flood attributes for varying levels of data and geographical factors in sub-Saharan Africa is also important. In addition, in order to address and mitigate the flood catastrophe and its associated impact in the sub-Saharan African region, local governments, and emergency organizations should consider the development of local maps and hazard-based land use management. The realization of sustainable strategies for efficient mitigation of the flood is necessary for the reduction of the impact and cost of economic losses on public health and safety. However, it should not limit the source of sound and existing knowledge and tools but promote, where possible, its use in urban planning, land use management, and sustainable development policies in the sub-Saharan African region. This paper has found that large percentages of residential locations in Sub-Saharan Africa are exposed to a significant level of flood hazard, and the region still lacks in studying and reducing its vulnerability. Recent attempts at mapping flood exposure in developing economies have relied on a passive approach to global flood monitoring. This approach is found to overestimate the temporally permanent distribution of flood hazard locations. As a result, residents are generally uninformed or, worse, misinformed of the actual hazards posed by their environment. This work also showed that signal detection and flood analyses can be used collectively with freely available optical datasets to mitigate the issues of previous methods. This methodology allowed the quick and systematic calculation of building-level flood exposure measures across the whole Sub-Saharan African states. Given the particular vulnerability of Sub-Saharan African societies to natural hazards, the percentage of flood hazard exposure could support improved strategies that situate both increased resilience and sustainability at a central role. Some essential policies that support this objective are emphasized with respect to early flood management and potential building fill costs. Such policies must balance the increased benefits to future communities associated with advanced hazard protection against the increased loss adjustment costs that enterprises will face when increasing coastal erosion hazard exposure is halted. Protecting at least landing-floor residential spaces located in these positions can prevent displacement and the accompanying strain on social housing resources. Public and private stakeholders, in turn, may gain increased interest in residential flood protection, which, in the absence of adaptive strategies, involves employing static interventions to prevent future flood impacts near the river coast. Huge resources are still specifically required, so information from flood exposure measures toward such low-lying areas is, therefore, crucial.

REFERENCES

Adelekan, I. O. (2020). Urban dynamics, everyday hazards and disaster risks in Ibadan, Nigeria. *Environment and Urbanization*, *32*(1), 213–232. Advance online publication. DOI: 10.1177/0956247819844738

Akinsemolu, A. A., Onyeaka, H. N., & Tamasiga, P. (2024). Climate-smart agriculture as a possible solution to mitigate climate change impact on food security in Sub-Saharan Africa. In *Food and Energy Security* (Vol. 13, Issue 1). DOI: 10.1002/fes3.509

Anami, S., De Block, M., MacHuka, J., & Van Lijsebettens, M. (2009). Molecular improvement of tropical maize for drought stress tolerance in Sub-Saharan Africa. *Critical Reviews in Plant Sciences*, *28*(1–2), 16–35. Advance online publication. DOI: 10.1080/07352680802665305

Bukari, C., & Aluko, O. A. (2023). Severity of climate change and deprivation outcomes: Micro-level assessment for sub-Saharan Africa. *Environmental Science & Policy*, *150*, 103593. Advance online publication. DOI: 10.1016/j.envsci.2023.103593

Chilunjika, A., & Gumede, N. (2021). Climate Change and Human Security in Sub-Saharan Africa. *African Renaissance, 2021*(si1). DOI: 10.31920/2516-5305/v2021sin1a2

Codjoe, S. N. A., & Atiglo, D. Y. (2020). The Implications of Extreme Weather Events for Attaining the Sustainable Development Goals in Sub-Saharan Africa. In *Frontiers in Climate* (Vol. 2). DOI: 10.3389/fclim.2020.592658

Cuthbert, M. O., Taylor, R. G., Favreau, G., Todd, M. C., Shamsudduha, M., Villholth, K. G., MacDonald, A. M., Scanlon, B. R., Kotchoni, D. O. V., Vouillamoz, J. M., Lawson, F. M. A., Adjomayi, P. A., Kashaigili, J., Seddon, D., Sorensen, J. P. R., Ebrahim, G. Y., Owor, M., Nyenje, P. M., Nazoumou, Y., & Kukuric, N. (2019). Observed controls on resilience of groundwater to climate variability in sub-Saharan Africa. *Nature*, *572*(7768), 230–234. Advance online publication. DOI: 10.1038/s41586-019-1441-7 PMID: 31391559

De Perez, E. C., Stephens, E., Bischiniotis, K., Van Aalst, M., Van Den Hurk, B., Mason, S., Nissan, H., & Pappenberger, F. (2017). Should seasonal rainfall forecasts be used for flood preparedness? *Hydrology and Earth System Sciences*, *21*(9), 4517–4524. Advance online publication. DOI: 10.5194/hess-21-4517-2017

Djègbè, I., Zinsou, M., Dovonou, E. F., Tchigossou, G., Sogbo, M., Adéoti, R., Gbaguidi, B., Atoyebi, S., Chandre, F., Akogbéto, M., Lines, J., & Djouaka, R. (2020). Minimal tillage and intermittent flooding farming systems show a potential reduction in the proliferation of Anopheles mosquito larvae in a rice field in Malanville, Northern Benin. *Malaria Journal, 19*(1), 333. Advance online publication. DOI: 10.1186/s12936-020-03406-2 PMID: 32928223

Dossou-Yovo, E. R., Devkota, K. P., Akpoti, K., Danvi, A., Duku, C., & Zwart, S. J. (2022). Thirty years of water management research for rice in sub-Saharan Africa: Achievement and perspectives. In *Field Crops Research* (Vol. 283). DOI: 10.1016/j.fcr.2022.108548

Dossou-Yovo, E. R., Kouadio, S. A. K., & Saito, K. (2023). Effects of mid-season drainage on iron toxicity, rice yield, and water productivity in irrigated systems in the derived savannah agroecological zone of West Africa. *Field Crops Research, 296*, 108901. Advance online publication. DOI: 10.1016/j.fcr.2023.108901

Echendu, A. J. (2021). Women, Development, and Flooding Disaster Research in Nigeria: A Scoping Review. *Environmental and Earth Sciences Research Journal, 8*(4), 147–152. Advance online publication. DOI: 10.18280/eesrj.080401

Gannon, K. E., Conway, D., Pardoe, J., Ndiyoi, M., Batisani, N., Odada, E., Olago, D., Opere, A., Kgosietsile, S., Nyambe, M., Omukuti, J., & Siderius, C. (2018). Business experience of floods and drought-related water and electricity supply disruption in three cities in sub-Saharan Africa during the 2015/2016 El Niño. *Global Sustainability, 1*, e14. Advance online publication. DOI: 10.1017/sus.2018.14

Geris, J., Comte, J. C., Franchi, F., Petros, A. K., Tirivarombo, S., Selepeng, A. T., & Villholth, K. G. (2022). Surface water-groundwater interactions and local land use control water quality impacts of extreme rainfall and flooding in a vulnerable semi-arid region of Sub-Saharan Africa. *Journal of Hydrology (Amsterdam), 609*, 127834. Advance online publication. DOI: 10.1016/j.jhydrol.2022.127834

Griffith, D. C., Kelly-Hope, L. A., & Miller, M. A. (2006). Review of reported cholera outbreaks worldwide, 1995-2005. In *American Journal of Tropical Medicine and Hygiene* (Vol. 75, Issue 5). DOI: 10.4269/ajtmh.2006.75.973

Gudoshava, M., Wanzala, M., Thompson, E., Mwesigwa, J., Endris, H. S., Segele, Z., Hirons, L., Kipkogei, O., Mumbua, C., Njoka, W., Baraibar, M., de Andrade, F., Woolnough, S., Atheru, Z., & Artan, G. (2022). Application of real time S2S forecasts over Eastern Africa in the co-production of climate services. *Climate Services, 27*, 100319. Advance online publication. DOI: 10.1016/j.cliser.2022.100319

Ibrahim, B., & Mensah, H. (2022). Rethinking climate migration in sub-Saharan Africa from the perspective of tripartite drivers of climate change. *SN Social Sciences*, *2*(6), 87. Advance online publication. DOI: 10.1007/s43545-022-00383-y

Jongman, B., Ward, P. J., & Aerts, J. C. J. H. (2012). Global exposure to river and coastal flooding: Long term trends and changes. *Global Environmental Change*, *22*(4), 823–835. Advance online publication. DOI: 10.1016/j.gloenvcha.2012.07.004

Joshua, M. D., Tompkins, E., Schreckenberg, K., Ngongondo, C., Gondwe, E., & Chiotha, S. (2022). Water policy and resilience of potable water infrastructure to climate risks in rural Malawi. *Physics and Chemistry of the Earth Parts A/B/C*, *127*, 103155. Advance online publication. DOI: 10.1016/j.pce.2022.103155

Leauthaud, C., Duvail, S., Hamerlynck, O., Paul, J. L., Cochet, H., Nyunja, J., Albergel, J., & Grünberger, O. (2013). Floods and livelihoods: The impact of changing water resources on wetland agro-ecological production systems in the Tana River Delta, Kenya. *Global Environmental Change*, *23*(1). Advance online publication. DOI: 10.1016/j.gloenvcha.2012.09.003

Lemaire, X. (2021). Climate change and strategic low-carbon planning in African cities after COVID-19: Inclusiveness or chaos? *Journal of the British Academy*, *9*, 39–79. Advance online publication. DOI: 10.5871/jba/009s9.039

Li, C., Yu, W., Dzodzomenyo, M., Asamoah, M., Kerapetse, C. T., Kandel, M., & Wright, J. (2021). Growing Spatial Overlap Between Dam-Related Flooding, Cropland and Domestic Water Points: A Water–Energy–Food Nexus Management Challenge in Malawi and Ghana. *Frontiers in Water*, *3*, 730370. Advance online publication. DOI: 10.3389/frwa.2021.730370

Mangani, R., Tesfamariam, E. H., Bellocchi, G., & Hassen, A. (2018). Growth, development, leaf gaseous exchange, and grain yield response of maize cultivars to drought and flooding stress. *Sustainability (Basel)*, *10*(10), 3492. Advance online publication. DOI: 10.3390/su10103492

Møller-Jensen, L., Agergaard, J., Andreasen, M. H., Kofie, R. Y., Yiran, G. A. B., & Oteng-Ababio, M. (2023). Probing political paradox: Urban expansion, floods risk vulnerability and social justice in urban Africa. *Journal of Urban Affairs*, *45*(3), 505–521. Advance online publication. DOI: 10.1080/07352166.2022.2108436

Mubangizi, A., Wanyama, J., Kiggundu, N., & Nakawuka, P. (2023). Assessing Suitability of Irrigation Scheduling Decision Support Systems for Lowland Rice Farmers in Sub-Saharan Africa—A Review. *Agricultural Sciences*, *14*(02), 219–239. Advance online publication. DOI: 10.4236/as.2023.142015

Müller, C., Ouédraogo, W. A., Schwarz, M., Barteit, S., & Sauerborn, R. (2023). The effects of climate change-induced flooding on harvest failure in Burkina Faso: Case study. *Frontiers in Public Health*, *11*, 1166913. Advance online publication. DOI: 10.3389/fpubh.2023.1166913 PMID: 37614457

Müller, C., Waha, K., Bondeau, A., & Heinke, J. (2014). Hotspots of climate change impacts in sub-Saharan Africa and implications for adaptation and development. *Global Change Biology*, *20*(8), 2505–2517. Advance online publication. DOI: 10.1111/gcb.12586 PMID: 24796720

Nagata, J. M., Hampshire, K., Epstein, A., Lin, F., Zakaras, J., Murnane, P., Charlebois, E. D., Tsai, A. C., Nash, D., & Weiser, S. D. (2022). Analysis of Heavy Rainfall in Sub-Saharan Africa and HIV Transmission Risk, HIV Prevalence, and Sexually Transmitted Infections, 2005-2017. *JAMA Network Open*, *5*(9), e2230282. Advance online publication. DOI: 10.1001/jamanetworkopen.2022.30282 PMID: 36074468

Negev, M., Teschner, N., Rosenthal, A., Levine, H., Lew-Levy, C., & Davidovitch, N. (2019). Adaptation of health systems to climate-related migration in Sub-Saharan Africa: Closing the gap. *International Journal of Hygiene and Environmental Health*, *222*(2), 311–314. Advance online publication. DOI: 10.1016/j.ijheh.2018.10.004 PMID: 30503929

Neumann, B., Vafeidis, A. T., Zimmermann, J., & Nicholls, R. J. (2015). Future coastal population growth and exposure to sea-level rise and coastal flooding - A global assessment. *PLoS One*, *10*(3), e0118571. Advance online publication. DOI: 10.1371/journal.pone.0118571 PMID: 25760037

Njoku, N., Lamond, J., Everett, G., & Manu, P. (2015). An overview of municipal solid waste management in developing and developed economies: Analysis of practices and contributions to urban flooding in Sub-Saharan Africa. *12th International Postgraduate Research Conference Proceedings*.

Perez-Saez, J., Lessler, J., Lee, E. C., Luquero, F. J., Malembaka, E. B., Finger, F., Langa, J. P., Yennan, S., Zaitchik, B., & Azman, A. S. (2022). The seasonality of cholera in sub-Saharan Africa: A statistical modelling study. *The Lancet. Global Health*, *10*(6), e831–e839. Advance online publication. DOI: 10.1016/S2214-109X(22)00007-9 PMID: 35461521

Ramiaramanana, F. N., & Teller, J. (2021). Urbanization and floods in sub-saharan africa: Spatiotemporal study and analysis of vulnerability factors—case of antananarivo agglomeration (Madagascar). *Water (Basel)*, *13*(2), 149. Advance online publication. DOI: 10.3390/w13020149

Rentschler, J., Salhab, M., & Jafino, B. A. (2022). Flood exposure and poverty in 188 countries. *Nature Communications, 13*(1), 3527. Advance online publication. DOI: 10.1038/s41467-022-30727-4 PMID: 35764617

Rother, H. A., Hayward, R. A., Paulson, J. A., Etzel, R. A., Shelton, M., & Theron, L. C. (2022). Impact of extreme weather events on Sub-Saharan African child and adolescent mental health: The implications of a systematic review of sparse research findings . In *Journal of Climate Change and Health* (Vol. 5). DOI: 10.1016/j.joclim.2021.100087

Serdeczny, O., Adams, S., Baarsch, F., Coumou, D., Robinson, A., Hare, W., Schaeffer, M., Perrette, M., & Reinhardt, J. (2017). Climate change impacts in Sub-Saharan Africa: From physical changes to their social repercussions. *Regional Environmental Change, 17*(6), 1585–1600. Advance online publication. DOI: 10.1007/s10113-015-0910-2

Shafik, W. (2024a). Environmental Conservation and Sustainable Development. In *Developing Digital Inclusion Through Globalization and Digitalization* (pp. 142–176). IGI Global. DOI: 10.4018/979-8-3693-4111-7.ch008

Shafik, W. (2024b). Navigating Emerging Challenges in Robotics and Artificial Intelligence in Africa. In *Examining the rapid advance of digital technology in Africa* (pp. 126–146). IGI Global., DOI: 10.4018/978-1-6684-9962-7.ch007

Shafik, W. (2024c). Sustainable Development. In Viana Hassan, A. S. S. B. (Ed.), *Building Community Resiliency and Sustainability With Tourism Development* (1st ed., Vol. 1, pp. 1–30). IGI Global., DOI: 10.4018/979-8-3693-5405-6.ch001

Shafik, W. (2024d). Sustainable development goal 14: explainable AI (XAI) for ocean health. In *Artificial Intelligence and Edge Computing for Sustainable Ocean Health* (pp. 167–198). Springer.

Shafik, W., & Kalinaki, K. (2024). Societal and Ethical Implications of Technology-Enhanced Agriculture and Healthcare: An African Context. *2024 IST-Africa Conference (IST-Africa)*, 1–11. DOI: 10.23919/IST-Africa63983.2024.10569306

Shah, H. A., Carrasco, L. R., Hamlet, A., & Murray, K. A. (2022). Exploring agricultural land-use and childhood malaria associations in sub-Saharan Africa. *Scientific Reports, 12*(1), 4124. Advance online publication. DOI: 10.1038/s41598-022-07837-6 PMID: 35260722

Simatele, D., Binns, T., & Simatele, M. (2012). Sustaining livelihoods under a changing climate: The case of urban agriculture in Lusaka, Zambia. *Journal of Environmental Planning and Management*, *55*(9), 1175–1191. Advance online publication. DOI: 10.1080/09640568.2011.637688

Tiepolo, M., & Galligari, A. (2021). Urban expansion-flood damage nexus: Evidence from the Dosso Region, Niger. *Land Use Policy*, *108*, 105547. Advance online publication. DOI: 10.1016/j.landusepol.2021.105547

Uttajug, A., Ueda, K., Seposo, X., & Francis, J. M. (2023). Association between extreme rainfall and acute respiratory infection among children under-5 years in sub-Saharan Africa: An analysis of Demographic and Health Survey data, 2006-2020. *BMJ Open*, *13*(4), e071874. Advance online publication. DOI: 10.1136/bmjopen-2023-071874 PMID: 37185183

Chapter 2
Examining Purchase and Usage Behavior Towards Arecanut Products Using Theory of Planned Behavior

K. S. Ganesha
https://orcid.org/0000-0001-8181-2411
Dayananda Sagar College of Engineering, India

Sharath Shridhar Hegde
Dayananda Sagar College of Engineering, India

ABSTRACT

The study focuses on purchase and usage behavior towards arecanut products using the theory of planned behavior. As we are aware, arecanut is a traditional product consumed by people in India. The objective of the study is to identify the factors affecting purchase and usage behavior towards arecanut products and the effect of purchase attitude, subjective norm, perceived behavior control, intention to buy, satisfaction, and facilities on purchase and usage behavior. The research is exploratory in nature because it tries to explore purchase and usage behavior towards arecanut products. The study shows factors such as purchase attitude, subjective norm, intention to buy, and consumer behavior affecting the purchase and usage behavior of arecanut products. Through analysis, it is found that there is an impact of purchase attitude, subjective norm, intention to buy, and consumer behavior on the purchase and usage behavior of arecanut products.

DOI: 10.4018/979-8-3373-3206-2.ch002

1. INTRODUCTION

The study focuses on purchase and usage behavior towards arecanut products using theory of planned behavior. As we aware arecanut is a traditional product which consumed by people of India. Arecanut is an important product which is consumed and used as a product, for religious purpose since many centuries as a tradition. However, in the present day there are lot of innovation with respect to arecanut products such as arecanut tee arecanut pan masala.

Since arecanut is a traditionally associated product, the promotion and branding of these products are very important. The organizations can use lot of latest promotional channels such as social media, websites. This will create awareness and interest about health benefits and usage of arecanut products to wide variety of customers. To purchase products like this, customers always plan and purchase. There are many factors which effect the purchase and usage behavior of arecanut products. So, to understand this we have used theory of planed behavior as a base model for the study.

2. LITERATURE REVIEW

Arecanut Products Usage

Arecanut is a product which is replacing tobacco in some countries. (Gupta, P. C., & Warnakulasuriya, S. 2002). However, there are people to use arecanut with or without tobacco also, which extends usage of arecanut. (Croucher, R., & Islam, S. 2002). Also, the usage of arecanut will create the addiction. (Benegal, V., Rajkumar, R. P., & Muralidharan, K. 2008). Irrespective of all these results the study of (Gunaseelan, R., Sowmya, R., & Sankaralingam, S., 2007), Shows that there is increasing arecanut usage, supporting to this Mehrtash, et. al., (2017), study also proves that there is increasing usage of arecanut.

Theory of Planned Behavior

The study shows there are several factors effect theory of planned behavior. It is important to note that factors such as norms, self-belief, and affect theory of planned behavior (Conner, M., & Armitage, C. J. 1998). The studies of (Bosnjak, M., Ajzen, I., & Schmidt, P. 2020). There is opportunity to extend theory of planned behavior. The past experiences and behavior of individuals are related to each other regarding to theory of planned behavior (Ajzen, I., 1991). So, it creates need for applying this theory.

Purchase Attitude: The study purchase and usage behavior knowing the attitude is the most important factor (Spears, N., & Singh, S. N., 2004). Also, the purchase attitude of the consumers affects the post purchase attitude (Geva, A., & Goldman, A., 1991). However, attitude always affects decisions of consumer (Tan, B. C., & Lau, T. C., 2011). When it comes to food products there is positive relationship between purchase and attitude (Huang, L., Bai, L., Zhang, X., & Gong, S., 2019).

Subjective Norm: The role of subjective norm is inconclusive (Schepers, J., & Wetzels, M., 2007). However subjective norm can be measured both directly and indirectly (De Vries, H., Dijkstra, M., & Kuhlman, P., 1988). It is very complex to manipulate subjective norms of the individuals (Trafimow, D., & Fishbein, M., 1994). Subjective norm and persevered behavior control will complement each other. (La Barbera, F., & Ajzen, I., 2020).

Perceived behavior control: Perceived behavior and perceiver control together forms perceived behavior control (Trafimow, D., Sheeran, P., Conner, M., & Finlay, K. A., 2002). Also Perceived ease or perceived difficulty of making a purchase decision also creates perceived behavior control (Ajzen, I., 2002) However perceived behavior control intention to buy the product. (Ajzen, I., & Madden, T. J., 1986). However, studies of Godin, G., Valois, P., & Lepage, L., (1993). Argues that perceived behavior control affects intention to exercise a purchase behavior but not prediction of existing behavior.

Intention to Buy: Intention to buy increases the purchase probability (Bemmaor, A. C., 1995). Supporting to this intention to buy increases purchase of intention to buy (Hajli, N., 2015). Purchase intentions of consumers towards personalized products will affects more towards purchase decisions. (Moon, J., Chadee, D., & Tikoo, S., 2008). Social media and modern communications affect intention to buy any products or services (Ismagilova, E., Slade, E. L., Rana, N. P., & Dwivedi, Y. K., 2020).

From the various literature review it is found that there are many studies which used theory of planned behavior. However, there are very few studies on purchase and usage of theory of planned behavior.

3. METHODS AND MATERIALS

The research is exploration in nature exploratory because it tries to explore purchase and usage behavior towards arecanut products. The study collected 103 responses from Indian consumers and users of arecanut. Data was collected through a structured questionnaire. Data was collected through google form by sharing its link to customers.

The questionnaire consisted of various demographic 34 statements like purchase attitude, Subjective norm, intention to buy, consumer behavior measured through Likert scale. The study justified sample size by taking factor analysis as a base, as per studies 1:3 is minimum sample ratio for factor analysis, as we have 34 statements, the sample size makes to 102.

4. Data ANALYSIS AND RESULTS

Table 1. Purchasing of arecanut from

	Frequency	**Percent**
Local Vendor	59	57.3
Mall	10	9.7
Online	2	1.9
Super Market	32	31.1
Total	103	100.0

Most of the consumers are Purchasing of arecanut from local vendor.

Table 2. Arecanut spending in rupees per month

	Frequency	*Percent*
< 500	4	3.9
1001 - 2,500	66	64.1
500 - 1,000	33	32.0
Total	103	100.0

Every month the peoples are spending 1001 - 2,500 rupees for arecanut products.

Factor Analysis

Table 3. KMO and Bartlett's test

Kaiser-Meyer-Olkin Measure of Sampling Adequacy.		**0.723**
Bartlett's Test of Sphericity	Approx. Chi-Square	2550.542
	df	231
	Sig.	.000

The range of 0 and 1, with higher qualities showing better reasonableness for factor examination. A KMO esteem above 0.6 is by and large viewed as satisfactory, while values above 0.8 are viewed as generally excellent.

Table 4. Total variance explained

Component	**Initial Eigenvalues**			**Extraction Sums of Squared Loadings**			**Rotation Sums of Squared Loadings**		
	Total	% of Variance	Cumulative %	Total	% of Variance	Cumulative %	Total	% of Variance	Cumulative %
1	9.648	43.855	43.855	9.648	43.855	43.855	3.466	38.483	38.483
2	2.654	12.064	55.919	2.654	12.064	55.919	2.514	11.429	49.913
3	2.196	9.981	65.900	2.196	9.981	65.900	2.029	9.223	59.136
4	1.763	8.015	73.915	1.763	8.015	73.915	2.002	9.099	68.235
5	1.192	5.420	79.335	1.192	5.420	79.335	1.971	8.958	77.192
6	1.058	4.807	84.142	1.058	4.807	84.142	1.529	6.950	84.142
7	.658	2.991	87.133						
8	.533	2.425	89.558						
9	.496	2.253	91.811						
10	.384	1.745	93.556						
11	.271	1.230	94.785						
12	.225	1.022	95.808						
13	.186	.844	96.652						
14	.181	.822	97.473						
15	.140	.635	98.109						
16	.117	.534	98.643						
17	.090	.411	99.054						
18	.084	.380	99.434						
19	.042	.193	99.627						

continued on following page

Table 4. Continued

Component	Initial Eigenvalues			Extraction Sums of Squared Loadings			Rotation Sums of Squared Loadings		
20	.036	.164	99.792						
21	.029	.134	99.926						
22	.016	.074	100.000						

Table 5. Factor loadings and reliability

Factor	Purchase Attitude	Subjective Norm	Perceived behavior control	Intention to Buy	Satisfaction	Facilities
I buy arecanut, whenever I like to buy	.909					
I do not get fresh arecanut throughout the year	.907					
I have knowledge on how to evaluate arecanut quality	.901					
I plan to buy arecanut	.879					
People in my community consume arecanut regularly	.871					
I have control in buying arecanut	.869					
I prefer arecanut that is easy and quick to use	.832					
I can buy arecanut confidently	.819					
I recommend my regular seller to my friends	.784					
I buy arecanut from regular seller	.756					
arecanut consumption is my part of culture	.644					
How do you rate Co-Operation Of staff during service?		.814				
People in my community consume arecanut regularly		.788				

continued on following page

Table 5. Continued

Factor	Purchase Attitude	Subjective Norm	Perceived behavior control	Intention to Buy	Satisfaction	Facilities
Are you satisfied One stop shop for Customers' needs.		.668				
My arecanut spend in Rupees per month is			.852			
I buy arecanut from			.590			
I feel arecanut is expensive				.865		
Buying arecanut takes a lot of time				.714		
Are you satisfied the platform to Purchase and Sale of arecanut and related Agriculture products?					.901	
How do you Rate of Totgars Cooperative sale society variety Of Product ?					.871	
How do you Rate of Store Facility?						.800
Technology helps me to buy arecanut						-.614
Cronbach Alpha	0.955	0.310	0.081	0.725	0.177	0.549
Variance %	43.855	12.064	9.981	8.015	5.420	4.807
Cumulative Variance	43.855	55.919	65.899	73.914	79.335	84.142

Factor 1 (Purchase Attitude): It explains the highest amount of variance (43.855%)
Factor 2 (Subjective Norm): It explains 12.064% of the variance and has a relatively lower Cronbach's alpha (0.310)
Factor 3 (Perceived Behavior Control): This factor has low loadings on variables related to perceived behavior control, explaining 9.981% of the variance
Factor 4 (Intention to Buy): This factor related to intention to buy arecanut, explaining 8.015% of the variance. The Cronbach's alpha for this factor is 0.725
Factor 5 (Satisfaction): This factor explaining 5.420% of the variance. The Cronbach's alpha for this factor is relatively low (0.177)
Factor 6 (Facilities): This factor explaining 4.807% of the variance. The Cronbach's alpha for this factor is low (0.549)

Regression

Table 6. Descriptive statistics

	Mean	Std. Deviation	N
PUB2	71.3495	13.78953	103
f1	38.5243	10.90138	103
f4	6.8932	2.55494	103
f6	5.7087	1.31828	103

The Mean is the average value of the data points in each variable. It represents the central tendency of the data and highest is PUB2.

The Standard Deviation measures the dispersion or spread of the data points around the Mean and highest is PUB2.

Table 7. Model summary

Model	R	R Square	Adjusted R Square	Std. Error of the Estimate	Change Statistics					Durbin-Watson
					R Square Change	F Change	df1	df2	Sig. F Change	
1	.944[a]	.890	.889	4.59125	.890	819.106	1	101	.000	
2	.961[b]	.923	.922	3.85436	.033	43.310	1	100	.000	
3	.978[c]	.956	.954	2.94877	.032	71.853	1	99	.000	1.732

a. Predictors: (Constant), f1

b. Predictors: (Constant), f1, f4

c. Predictors: (Constant), f1, f4, f6

d. Dependent Variable: PUB2

The correlation coefficient between the observed and predicted values is 0.944. This indicates a strong positive linear relationship between the predictors and the dependent variable PUB2.

Table 8. ANOVA

Model		Sum of Squares	df	Mean Square	F	Sig.
1	Regression	17266.384	1	17266.384	819.106	.000[b]
	Residual	2129.034	101	21.080		
	Total	19395.417	102			
2	Regression	17909.806	2	8954.903	602.775	.000[c]
	Residual	1485.612	100	14.856		
	Total	19395.417	102			
3	Regression	18534.587	3	6178.196	710.525	.000[d]
	Residual	860.830	99	8.695		
	Total	19395.417	102			

a. Dependent Variable: PUB2
b. Predictors: (Constant), f1

In all three ANOVA tests, the values are extremely small, well below the significance level of 0.05

Table 9. Regression coefficients

Model		Unstandardized Coefficients		Standardized Coefficients	t	Sig.	95.0% Confidence Interval for B		Correlations			Collinearity Statistics	
		B	Std. Error	Beta			Lower Bound	Upper Bound	Zero-order	Partial	Part	Tolerance	VIF
1	(Constant)	25.371	1.669		15.201	.000	22.060	28.682					
	f1	1.193	.042	.944	28.620	.000	1.111	1.276	.944	.944	.944	1.000	1.000
2	(Constant)	18.848	1.716		10.981	.000	15.443	22.253					
	f1	1.187	.035	.938	33.888	.000	1.117	1.256	.944	.959	.938	.999	1.001
	f4	.983	.149	.182	6.581	.000	.687	1.280	.209	.550	.182	.999	1.001
3	(Constant)	3.485	2.238		1.557	.123	-.956	7.926					
	f1	1.270	.029	1.004	44.510	.000	1.213	1.327	.944	.976	.942	.881	1.135
	f4	1.083	.115	.201	9.420	.000	.855	1.311	.209	.588	.199	.989	1.011
	f6	2.010	.237	.192	8.477	.000	1.540	2.481	-.175	.549	.179	.872	1.147

Overall, the coefficients in each model show the magnitude and direction of the relationships between the predictors and the dependent variable. Tolerance and VIF are well within the range above 0.25. So, there is no existence of multi-collinearity and auto correlation.

5. DISCUSSIONS

The study's findings provide valuable insights into the factors that influence the purchase and usage behavior of arecanut products. The results of the regression analysis reveal that purchase attitude, subjective norm, intention to buy, and consumer behavior all have a significant impact on the purchase and usage behavior of arecanut products.

Specifically, the study found that:

- Purchase attitude plays a crucial role in shaping consumer behavior, suggesting that consumers' positive or negative attitudes towards arecanut products significantly influence their purchasing decisions.
- Subjective norm, which refers to the perceived social pressure to conform to societal norms, also emerged as a significant predictor of purchase and usage behavior. This implies that consumers are more likely to purchase and use arecanut products if they perceive that others around them approve of or use these products.
- Intention to buy, which captures consumers' planned behavior, was found to be a strong predictor of actual purchase and usage behavior. This highlights the importance of understanding consumers' purchase intentions in predicting their behavior.
- Consumer behavior, which encompasses various psychological, social, and cultural factors, also emerged as a significant predictor of purchase and usage behavior. This suggests that consumers' individual characteristics, lifestyle, and cultural background all play a role in shaping their purchasing decisions.

Overall, the study's findings have important implications for marketers, policy-makers, and other stakeholders interested in promoting the sale and consumption of arecanut products. By understanding the factors that drive consumer behavior, these stakeholders can develop targeted strategies to promote positive attitudes, social norms, and purchase intentions towards arecanut products.

6. CONCLUSION

The marketers should focus on factors such as purchase attitude, Subjective norm, intention to buy, consumer behavior affecting the purchase and usage behavior of arecanut products. Ther results of the study are also matches with several studies and show that purchase and usage intention affect purchase behaviors. Also, we can conclude that theory of planned behavior is a tested theory for studying purchase and usage intention of consumers. From the study we can say it can be applied for even to products like arecanut.

7. IMPLICATIONS AND FUTURE SCOPE OF STUDY

The research under taken by the researcher would help the arecanut manufacturers and marketers to know about the purchase and usage intention of customers. The marketers should focus on factors such as purchase attitude, Subjective norm, intention to buy, consumer behavior affecting the purchase and usage behavior of arecanut products. The marketers can promote arecanut products based on these factors.

REFERENCES

Ajzen, I. (1991). The theory of planned behavior. *Organizational Behavior and Human Decision Processes, 50*(2), 179–211. DOI: 10.1016/0749-5978(91)90020-T

Ajzen, I. (2002). Perceived behavioral control, self-efficacy, locus of control, and the theory of planned behavior 1. *Journal of Applied Social Psychology, 32*(4), 665–683. DOI: 10.1111/j.1559-1816.2002.tb00236.x

Ajzen, I., & Madden, T. J. (1986). Prediction of goal-directed behavior: Attitudes, intentions, and perceived behavioral control. *Journal of Experimental Social Psychology, 22*(5), 453–474. DOI: 10.1016/0022-1031(86)90045-4

Bemmaor, A. C. (1995). Predicting behavior from intention-to-buy measures: The parametric case. *JMR, Journal of Marketing Research, 32*(2), 176–191. DOI: 10.1177/002224379503200205

Benegal, V., Rajkumar, R. P., & Muralidharan, K. (2008). Does areca nut use lead to dependence? *Drug and Alcohol Dependence, 97*(1-2), 114–121. DOI: 10.1016/j.drugalcdep.2008.03.016 PMID: 18490113

Bosnjak, M., Ajzen, I., & Schmidt, P. (2020). The theory of planned behavior: Selected recent advances and applications. *Europe's Journal of Psychology, 16*(3), 352–356. DOI: 10.5964/ejop.v16i3.3107 PMID: 33680187

Conner, M., & Armitage, C. J. (1998). Extending the theory of planned behavior: A review and avenues for further research. *Journal of Applied Social Psychology, 28*(15), 1429–1464. DOI: 10.1111/j.1559-1816.1998.tb01685.x

Croucher, R., & Islam, S. (2002). Socio-economic aspects of areca nut use. *Addiction Biology, 7*(1), 139–146. DOI: 10.1080/13556210120091518 PMID: 11900634

De Vries, H., Dijkstra, M., & Kuhlman, P. (1988). Self-efficacy: The third factor besides attitude and subjective norm as a predictor of behavioural intentions. *Health Education Research, 3*(3), 273–282. DOI: 10.1093/her/3.3.273

Geva, A., & Goldman, A. (1991). Satisfaction measurement in guided tours. *Annals of Tourism Research, 18*(2), 177–185. DOI: 10.1016/0160-7383(91)90002-S

Godin, G., Valois, P., & Lepage, L. (1993). The pattern of influence of perceived behavioral control upon exercising behavior: An application of Ajzen's theory of planned behavior. *Journal of Behavioral Medicine, 16*(1), 81–102. DOI: 10.1007/BF00844756 PMID: 8433359

Gunaseelan, R., Shanthi, S., Sowmya, R., & Datta, M. (2007). Areca nut use among rural residents of Sriperambudur Taluk: A qualitative study. *Indian Journal of Dental Research*, *18*(1), 11–14. DOI: 10.4103/0970-9290.30915 PMID: 17347538

Gupta, P. C., & Warnakulasuriya, S. (2002). Global epidemiology of areca nut usage. *Addiction Biology*, *7*(1), 77–83. DOI: 10.1080/13556210020091437 PMID: 11900626

Hajli, N. (2015). Social commerce constructs and consumer's intention to buy. *International Journal of Information Management*, *35*(2), 183–191. DOI: 10.1016/j.ijinfomgt.2014.12.005

Huang, L., Bai, L., Zhang, X., & Gong, S. (2019). Re-understanding the antecedents of functional foods purchase: Mediating effect of purchase attitude and moderating effect of food neophobia. *Food Quality and Preference*, *73*, 266–275. DOI: 10.1016/j.foodqual.2018.11.001

Ismagilova, E., Slade, E. L., Rana, N. P., & Dwivedi, Y. K. (2020). The effect of electronic word of mouth communications on intention to buy: A meta-analysis. *Information Systems Frontiers*, *22*(5), 1203–1226. DOI: 10.1007/s10796-019-09924-y

La Barbera, F., & Ajzen, I. (2020). Control interactions in the theory of planned behavior: Rethinking the role of subjective norm. *Europe's Journal of Psychology*, *16*(3), 401–417. DOI: 10.5964/ejop.v16i3.2056 PMID: 33680190

Mehrtash, H., Duncan, K., Parascandola, M., David, A., Gritz, E. R., Gupta, P. C., Mehrotra, R., Amer Nordin, A. S., Pearlman, P. C., Warnakulasuriya, S., Wen, C.-P., Zain, R. B., & Trimble, E. L. (2017). Defining a global research and policy agenda for betel quid and areca nut. *The Lancet. Oncology*, *18*(12), e767–e775. DOI: 10.1016/S1470-2045(17)30460-6 PMID: 29208442

Moon, J., Chadee, D., & Tikoo, S. (2008). Culture, product type, and price influences on consumer purchase intention to buy personalized products online. *Journal of Business Research*, *61*(1), 31–39. DOI: 10.1016/j.jbusres.2006.05.012

Schepers, J., & Wetzels, M. (2007). A meta-analysis of the technology acceptance model: Investigating subjective norm and moderation effects. *Information & Management*, *44*(1), 90–103. DOI: 10.1016/j.im.2006.10.007

Spears, N., & Singh, S. N. (2004). Measuring attitude toward the brand and purchase intentions. *Journal of Current Issues and Research in Advertising*, *26*(2), 53–66. DOI: 10.1080/10641734.2004.10505164

Tan, B. C., & Lau, T. C. (2011). Green purchase behavior: Examining the influence of green environmental attitude, perceived consumer effectiveness and specific green purchase attitude. *Australian Journal of Basic and Applied Sciences*, *5*(8), 559–567.

Trafimow, D., & Fishbein, M. (1994). The moderating effect of behavior type on the subjective norm-behavior relationship. *The Journal of Social Psychology*, *134*(6), 755–763. DOI: 10.1080/00224545.1994.9923010

Trafimow, D., Sheeran, P., Conner, M., & Finlay, K. A. (2002). Evidence that perceived behavioural control is a multidimensional construct: Perceived control and perceived difficulty. *British Journal of Social Psychology*, *41*(1), 101–121. DOI: 10.1348/014466602165081 PMID: 11970777

Chapter 3
Artificial Intelligence and Machine Learning for Geodata Analysis

Shraddha Vitthal Shingne
https://orcid.org/0009-0009-5911-2860
Dr. D.Y. Patil Institute of Technology, Pune, India

Aditya Bhila Pawar
Dr. D.Y. Patil Institute of Technology, Pune, India

ABSTRACT

Artificial Intelligence and Machine Learning have revolutionized geospatial data analysis, allowing accurate insights from extensive satellite imagery datasets. Deep learning models, especially CNNs, allow automatic feature extraction for land-use classification, disaster evaluation, and environmental monitoring. Platforms such as PyTorch and TensorFlow, combined with specialized libraries, streamline implementation. These developments enhance decision-making in urban planning, climate change assessment, and resource management.

1. BACKGROUND

The innovations in Machine Learning (ML) and Artificial Intelligence (AI) in geospatial analysis have mostly been a function of the transformation in data collection, more processing power, and the arrival of the big data era (Zhang & Zhang, 2020). Over the past decade, geospatial data generation has doubled exponentially with the use of satellite imagery, along with the growth in the utilization of remote sensing technologies and the creation of Geographic Information Systems (GIS)

DOI: 10.4018/979-8-3373-3206-2.ch003

(Yao, Yang, & Lin, 2018). When considered collectively, all these generate an extremely large amount of data, which needs more advanced techniques to derive meaningful information, thereby leading to improved decision-making procedures in urban planning, environmental monitoring, and disaster recovery (Cheng, Han, & Lu, 2017).

Geospatial data analysis was once a task of interpretation by methods manually installed. These methods were time-consuming and prone to errors, where the strategy was centered more on extremely basic statistical and rule-based methods. This was largely dependent on human effort to pull out useful information from raw data (Srivastava, Pant, & Agarwal, 2020). Today, with AI and ML, the same processes have turned into automated analysis, lessening the range of human error and increasing efficiency.

Through the use of AI, deep learning, and neural network techniques, geospatial processing and interpretation have significantly changed. AI is able to identify patterns, land cover classification, change detection, and predict outcome through big datasets through the application of machine learning algorithms (Goodfellow, Bengio, & Courville, 2016). Deep learning models, especially Convolutional Neural Networks (CNNs), have been applied in unprecedented levels of land use classification, object detection, and feature extraction, leading to improved analysis of high-resolution satellite imagery (Ronneberger, Fischer, & Brox, 2015).

The powerful drivers of applying AI/ML in geospatial analysis are the following reasons:

1. **Technology Advancement:** The fast pace of technology development, with enormous leaps in computer power and vast storage capacities, coupled with the creation of sophisticated algorithms, has made it possible to apply AI-geospatial systems through various applications like Google Earth Engine and platforms like TensorFlow and PyTorch (He et al., 2016).
2. **Availability of Big Data:** With the large production and deployment of satellites and drones, big geospatial data have been created. These are multi-dimensional data by nature comprising spectral bands, time-series data, spatial resolution, or any other heterogeneous attributes of data that are best to be analyzed by methods associated with machine learning (Zhang et al., 2020). Through the inference of meaningful patterns from these attributes, which would otherwise remain undetected, artificial intelligence methods can be utilized optimally on the dataset itself.
3. **Real-Time Analytics:** Real-time analytics are necessary in certain sectors, like disaster response and environmental monitoring. Machine Learning and Artificial Intelligence can process real-time data gathered from diverse sources to give timely analytics and insights for timely decision support in flood management,

forest fire occurrences, and infrastructure development progress (Cheng et al., 2017).

4. **Precision and Accuracy:** The conventional geospatial analytical methods tend to display extremely tolerant requirements in problem-solving in terms of precision and accuracy, particularly for more complex topographies. Using artificial intelligence variables reduces human error considerably, thus enhancing the overall precision of most operations in terms of classification or prediction. The continuous advancement of technological innovations, as well as the growing need for high-efficiency data-processing systems, drives the accelerated evolution of artificial intelligence and in turn machine learning within geospatial analysis (Yao et al., 2018).

These technological advancements are forecasted to enable the development of smart urban spaces, guarantee the maximization of agricultural practices, and play a role towards efficient planning of disaster management strategies, thereby making cities sustainable and resilient. Artificial intelligence and machine learning are in their infancy stages of their potentiality and are expanding by exponential growth.

It is in the background of growing need for efficient processing instruments in artificial intelligence and machine learning that new technologies have grown exponentially in recent times. Just like with this, complexity in design results in creating systems more sophisticated and advanced towards achieving a core goal in the realm of more sophisticated urban planning, agricultural development, and disaster management, respectively. AI and ML are just germinating in their potential, yet they are being achieved at such an unprecedented magnitude. In return, this will transform the interpretation and utilization of geospatial data, as well as reshape the world itself that individuals subsequently perceive and interact with (Goodfellow et al., 2016).

2. INTRODUCTION

Geospatial data analysis has become increasingly important, thanks to the vast amounts of information generated by the Earth's varied topography, climate, and human activities (Zhang & Zhang, 2020). With the advancements in remote-sensing technologies, we can now capture high-resolution images and detailed maps of the Earth's surface, resulting in a remarkable surge in geospatial data (Yao, Yang, & Lin, 2018).

2.1 Relevance of Geospatial Data Analysis

Geospatial data analysis plays a critical role in understanding and managing the Earth's resources. Some of the key areas where geospatial data analysis is highly relevant include:

- **Environment Monitoring:** Geospatial data is used to monitor environmental changes, such as deforestation, urbanization, and climate change. For example, satellite imagery can track the loss of forest cover over time, enabling governments and organizations to take corrective actions (Cheng, Han, & Lu, 2017).
- **Disaster Management:** Geospatial data is essential for assessing damage after natural disasters like earthquakes, floods, and hurricanes. It helps in planning rehabilitation efforts and allocating resources effectively (Ronneberger, Fischer, & Brox, 2015).
- **Agricultural Management:** Geospatial data is used to monitor crop health, predict yields, and optimize irrigation. This data helps farmers make data-driven decisions to improve agricultural productivity and sustainability (Srivastava, Pant, & Agarwal, 2020).
- **Urban Planning:** Geospatial data aids in urban development by providing insights into land use patterns, infrastructure planning, and traffic management. It helps city planners make informed decisions about zoning, transportation networks, and resource allocation (He et al., 2016).

Traditional geospatial analysis methods were often cumbersome, time-consuming, and prone to human error. However, with the integration of AI and ML techniques, the accuracy and efficiency of geospatial data analysis have significantly improved. AI and ML automate the interpretation of large datasets, uncovering patterns and trends that would be difficult to detect manually (Goodfellow, Bengio, & Courville, 2016).

2.2 Overview of AI and ML in Geospatial Analysis

AI and ML are the most important tools that apply in the course of geospatial analysis. They assist in automating processes for interpreting large datasets, thereby exposing patterns and trends (Zhang et al., 2020).

- **Machine Learning Techniques:** These are techniques that are trained from past data so that prediction of new unseen data can be predicted. Some of the most widely used ML techniques implemented for geospatial analytics include supervised classification, unsupervised classification, regression analysis, and time series analysis (Yao et al., 2018).

- **Deep Learning Models:** Convolutional Neural Network or CNNs became extremely popular because it is possible to learn hierarchical features from images in an end-to-end fashion. The models gave their way to success in image classification, detection of objects, and pixel-wise segmentation (Goodfellow et al., 2016).

2.3 Evolution of Geospatial Analysis Using AI/ML

The approaches in AI and ML for geospatial analysis, starting from the most simple image processing techniques to the deep learning algorithms used today, started with very basic pixel-based classifications. These early methods heavily relied on pixel-based classifications, which proved independent in classifying every single pixel in an image. Those proved not capable of spatial variability and could not deal with complicated patterns (Cheng et al., 2017). Some of the applications include machine learning, particularly supervised learning techniques-behavioral modeling that can be trained upon labeled data and look for particular patterns to predict results afterward with new data. Geospatial analysis was shifted toward deep learning through Convolutional Neural Networks. CNNs learn spatial hierarchies from their data, making them automatically very effective at detecting and classifying objects in satellite imagery, significantly reducing the need for manual feature engineering and greatly improving land-use classification, object detection, and monitoring tasks with much improved accuracy (Ronneberger et al., 2015).

Summarized below are the major steps in the analysis of geospatial data, tools and technologies, and applications of AI and ML in diversified areas that include land cover classification and disaster assessment (Zhang & Zhang, 2020).

3. PHASES OF AI/ML FOR GEODATA ANALYSIS

AI/ML Phases for Geodata Analysis. The steps involved in AI/ML for geodata analysis include a host of the key phases that contribute to the correct interpretation and exploitation of geospatial data.

3.1 Data Acquisition

Data acquisition is the spine of any geospatial analysis project. Essentially, it corresponds to data collection via a lengthy list of sources: remote sensing satellites, aerial drones, LiDAR, ground-based sensors. Every source of data has its characteristics and resolution along with the advantage it offers.

- **Satellite Imagery:** This source of data is the most used because it offers wide coverage. Satellites provide Landsat, Sentinel, and MODIS images that contain inherent time domain information that will enable analysts to study changes over a specified time period. However, the effects of such atmospheric conditions as cloud cover negatively affect the visibility of details (Yao, Yang, & Lin, 2018).
- **Aerial Photography and UAVs:** The Drones are unmanned aerial vehicles that can facilitate flexibility in the form of high-resolution images that are easily achieved at lower altitudes. That said, such usage would be most appropriate for projects that have to be analyzed for details in places that could involve vegetation mapping or infrastructure monitoring (Zhang & Zhang, 2020).
- **Ground-Based Sensors:** These would collect local environmental information such as soil moisture, temperature, and humidity, which would be significant for agricultural use. Combining the data with the satellite imagery would thus give a much more comprehensive view of the study area (Goodfellow, Bengio, & Courville, 2016).

Table 1. Comparison of geospatial data sources

Source	Advantages	Limitations
Satellite Imagery	Wide coverage, historical data	Limited resolution, affected by weather
Aerial Photography	High resolution, flexibility in capture	Limited coverage area, costly
Ground-Based Sensors	Real-time data, high accuracy	Limited spatial coverage, maintenance required

3.2 Pre-Processing

Pre-processing is a critical step in geospatial data analysis, as it enhances the quality of the data and prepares it for further analysis. This phase involves several operations aimed at cleaning and preparing the data for feature extraction and modelling.

- **Critical Operations Pre-Processing:**

1. **Noise Reduction:** Disturbances that are unwanted about an image are removed. One such technique includes Gaussian filter and median filter, which is used to remove noise without disturbing the main features in an image (OpenCV Documentation, 2024).
2. **Binarization:** This process converts the image into a binary format, where the text is white against a black background. Techniques like Otsu's thresholding or adaptive thresholding are commonly used for this purpose (Ronneberger, Fischer, & Brox, 2015).
3. **Edge Detection:** Edge detection algorithms, such as the Canny edge detector, are used to enhance the edges of objects in the image. This helps in differentiating between different features, such as land cover types or infrastructure (He et al., 2016).
4. **Image Smoothing:** Smoothing filters are applied to create a uniform transition in the image, reducing sharpness and ensuring that the image is suitable for further analysis

Figure 1. Pre-processing steps in AI/ML for geodata analysis

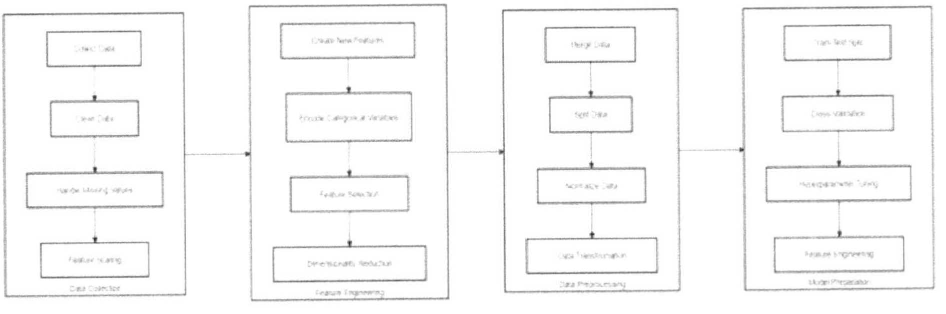

3.3 Feature Extraction

Feature extraction means it is simply retrieving some critical patterns in the data that may be taken for further examination. The raw data is converted into a structured format and then interpreted into representation with the help of machine learning models.

Best Techniques Used for Feature Extraction:

- **Scale-Invariant Feature Transform (SIFT):** This algorithm is actually focused on detection and description of local features in images. This makes the algorithm pretty effective for scenes where objects are scaled or rotated. This method proves especially useful in areas involving pattern recognition and terrain classification (Cheng, Han, & Lu, 2017).
- **Principal Component Analysis (PCA):** It reduces the dimension of the information of the data through transformation to a set of uncorrelated variables referred to as principal components. This simplification of information has the effect of speeding up the computation while retaining the core information (Zhang & Zhang, 2020).
- **Histogram of Oriented Gradients (HOG):** HOG centers its attention on the distribution of intensity gradients, which is valuable in the recognition of textures and shapes within an image. This makes it a very potent tool in the identification of land features in remote sensing (Nian, Suk, & Kim, 2013).

Table 2. Comparison of feature extraction techniques

Technique	*Description*	*Advantages*	*Limitations*
SIFT	Extracts distinctive features	Invariant to scale and rotation	Computationally expensive
PCA	Reduces dimensionality	Improves processing speed	May lose important features
HOG	Captures gradient information	Effective for character recognition	Sensitive to noise

3.4 Neural Networks

A CNN is a deep neural network that learns to progressively see features: simple edges and textures then more complex patterns. The typical architecture of a CNN contains the following components:

1. **Convolutional Layers:** These layers perform convolution operations using a number of filters/kernels, that learn to extract local patterns in the input image, which might look like simple edges, corners, and textures; each different kind of feature learned by a filter (Goodfellow, Bengio, & Courville, 2016).
2. **Pooling Layers:** After the convolutional layers are done, pooling layers reduce the size of feature maps without losing any of the highly important information. Probably one of the most popular methods of maximum-pooling is the one that selects the highest value in a designated region of the feature map (He et al., 2016).
3. **Fully Connected Layers**: These layers add all the features learnt from the preceding layers and decide the final classification decision. In the fully connected layer, each neuron of this layer is connected to each neuron of the preceding layer. This allows the network to sum all the features learnt towards the output (Goodfellow et al., 2016).
4. **Softmax Layer**: This layer computes the probabilities for the class labels and then it outputs the most probable class for that input image Apart from CNNs, other variants of neural networks might be used like Recurrent Neural Networks or LSTMs for character recognition in sequential data when the characters appear in a particular order, for example, cursive hand-written (Goodfellow, Bengio, & Courville, 2016).

Classification and recognition purposes have utilizations of neural networks in geospatial data analyses. They basically imitate the structure of a brain and are particularly strong dealing with complex input data.

Types of Neural Networks:

1. **Feedforward Neural Networks:** These networks consist of multiple layers of neurons, where information flows in one direction, from the input layer to the output layer. They are used for learning both linear and nonlinear relationships within the input data (Rojas, Arora, & Umscheid, 2023).

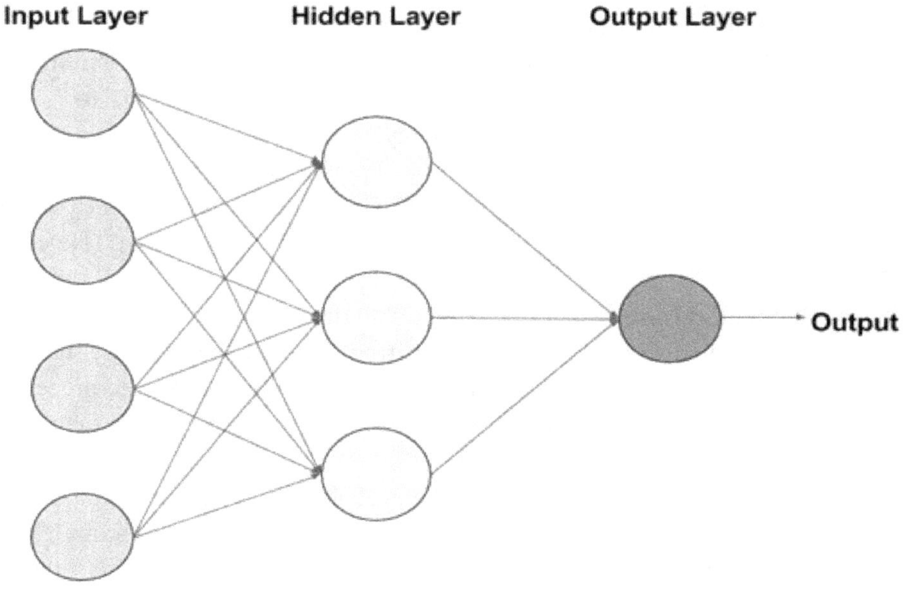

2. **Convolutional Neural Networks (CNNs):** CNNs are specifically designed for image-based tasks and are highly effective in automatically learning hierarchical features from data. They are widely used in geospatial analysis for tasks such as land cover classification, object detection, and image segmentation (Cheng, Han, & Lu, 2017; Zhang & Zhang, 2020).

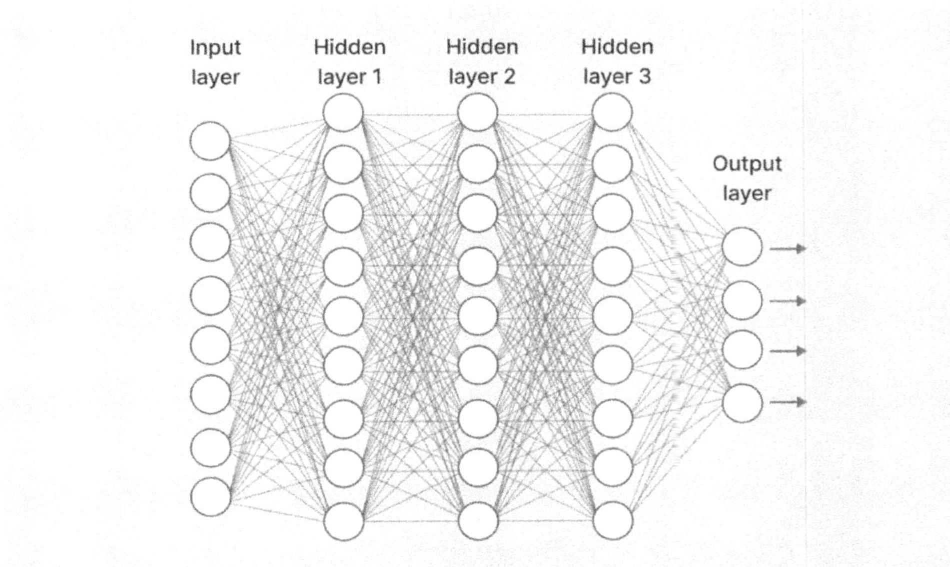
Input
layer
Hidden
layer 1
Hidden
layer 2
Hidden
layer 3
Output
layer

Figure 4. Typical architecture of a convolutional neural network (CNN)

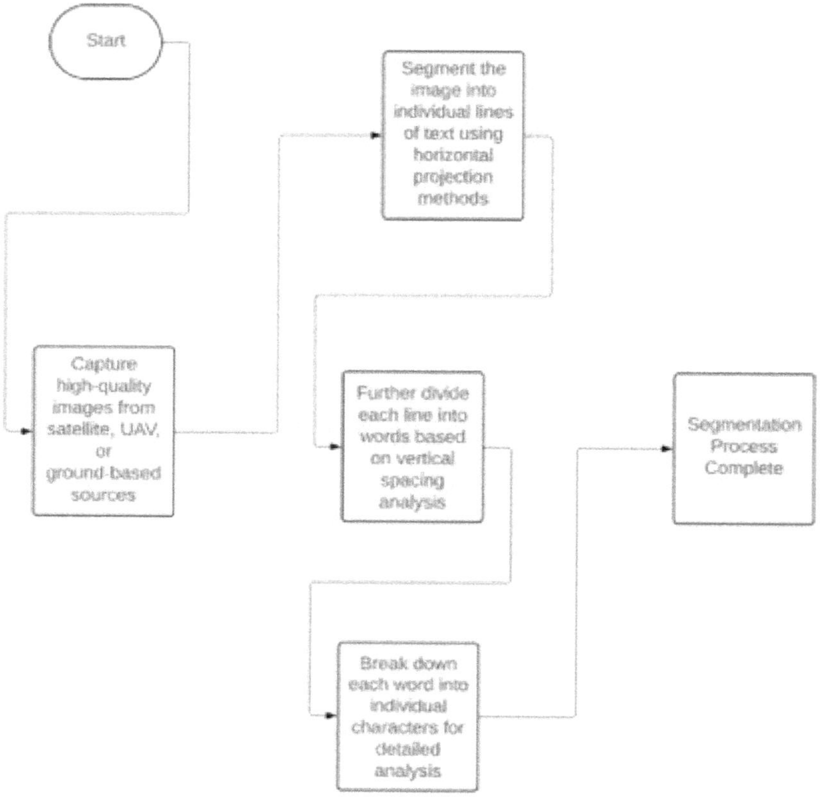

Applications of Neural Networks in Geospatial Analysis:

Neural networks, particularly CNNs, have been successfully applied in various geospatial analysis tasks, including:

i. **Land Cover Classification:** CNNs are used to classify land cover types, such as forests, urban areas, and water bodies, from satellite imagery. This information is crucial for urban planning, environmental monitoring, and resource management (Cheng, Han, & Lu, 2017).
ii. **Object Detection:** CNNs can detect specific objects, such as buildings, roads, and vehicles, in satellite images. This is particularly useful in disaster management, where rapid damage assessment is required (Mnih, Kavukcuoglu, & Silver, 2015).

iii. **Image Segmentation:** CNNs are used for pixel-wise segmentation of satellite images, enabling detailed analysis of land use patterns and changes over time (Ronneberger, Fischer, & Brox, 2015).

In summary, neural networks, particularly CNNs, have revolutionized geospatial data analysis by automating the extraction of features from complex datasets. Their ability to learn hierarchical patterns from images has significantly improved the accuracy and efficiency of tasks such as land cover classification, object detection, and image segmentation.

4. TOOLS AND TECHNOLOGIES

The tools and technologies used in geospatial data analysis significantly impact the efficiency and precision of data processing. A wide range of software libraries, frameworks, and platforms are available to support AI and ML applications in geospatial analysis. Below, we explore some of the most commonly used tools and technologies:

- **Python:** Python is one of the most widely used programming languages in geospatial data analysis due to its flexibility and extensive libraries. Libraries such as NumPy, Pandas, and SciPy provide robust support for data manipulation, analysis, and visualization. Additionally, geospatial-specific libraries like GDAL and Rasterio enable the processing of geospatial data formats, making Python an indispensable tool for professionals in this field (Python Software Foundation, 2023; NumPy, 2023; Pandas, 2023; GDAL, 2023; Rasterio, 2023).
- **TensorFlow and Keras:** TensorFlow is an open-source machine learning framework that provides a comprehensive ecosystem for building and deploying machine learning models. Keras, a high-level API that runs on top of TensorFlow, simplifies the creation and training of deep learning models. These frameworks are widely used for developing complex neural network architectures, particularly for image-based tasks in geospatial analysis (TensorFlow, 2023; Chollet, 2018).
- **Google Earth Engine (GEE):** GEE is a cloud-based platform that provides access to a vast repository of satellite imagery and geospatial datasets. It offers powerful tools for planetary-scale analysis and simplifies data processing through an easy-to-use code editor, available in both JavaScript and Python. GEE is particularly useful for large-scale geospatial projects, such as environmental monitoring and disaster management (Google Earth Engine, 2023).

- **OpenCV:** OpenCV is an open-source computer vision library that provides tools for image processing, filtering, and segmentation. It is widely used for preprocessing geospatial images, such as noise reduction, edge detection, and image smoothing, before feeding them into machine learning models (OpenCV, 2023).
- **NumPy:** NumPy is a fundamental package for scientific computing in Python. It provides support for large, multi-dimensional arrays and matrices, along with a collection of mathematical functions to operate on these arrays. NumPy is essential for numerical computations in geospatial data analysis (OpenCV, 2023).
- **Pandas:** Pandas is a library for data manipulation and analysis in Python. It provides data structures like DataFrames, which simplify the process of cleaning, transforming, and analyzing geospatial datasets. Pandas is particularly useful for preparing data for training machine learning models (Pandas, 2023).
- **QGIS:** QGIS is an open-source Geographic Information System (GIS) that supports the visualization, analysis, and editing of geospatial data. It provides a wide range of tools for working with vector and raster data, making it a popular choice for geospatial professionals (QGIS Development Team, 2023).

Table 2.

Sr No.	Tools	Technologies
1	**Python**	This language is widely used in geospatial analysis, with big libraries and ease of use. It supports such activities as data processing, automation, and working with databases.
2	**TensorFlow**	It is an open-source library for machine learning, a comprehensive ecosystem for building and deploying models, supporting both CPU and GPU computations..
3	**Keras**	A high-level neural network API which executes directly on top of TensorFlow, Keras streamlines the creation and training of deep learning models and is therefore very accessible to beginners and experts alike .
4	**OpenCV**	An open-source computer vision library, it provides tools for image processing, filtering and even segmentation. It's an essential preprocessing tool for geospatial images.
5	**NumPy**	It is the base package for any Python code for any type of scientific computing. For instance, large arrays and matrices can undergo operations, hence absolutely necessary for numerical calculations in the kind of data analysis that geospatial analysis requires.

continued on following page

Table 2. Continued

Sr No.	Tools	Technologies
6	**Pandas**	A library for data manipulation and analysis that provides data structures like DataFrames, which would be easier to manage in preparation for training models and testing them.
7	**Google Earth Engine (GEE)**	GEE is the natural offshoot of the blossoming cloud computing landscape- a cloud-based platform for planetary-scale geospatial analysis, with ready access to a vast repository of satellite imagery and geospatial datasets.

5. MOTIVATION

The rationale behind integrating AI and ML with geospatial data analysis is to analyze and bring meaning to the enormous volumes of data coming out from satellite images, remote sensing technologies, and GIS (Li et al., 2020; Zhang et al., 2019). Traditional methods of analyzing geospatial data are often deficient because they rely on manual processing and human interpretation, which require much time and are prone to human error (Goodchild & Li, 2012). Obviously, the advent of AI and ML changed geospatial analysis. What has emerged is a shift from rule-based methods to more data-driven methods (Reichstein et al., 2019). The automation provided by these technologies enables the delivery of processed information quicker, with greater accuracy and deeper insights into the data itself, thus enabling applications in such industries as urban planning, agriculture, disaster management, and environmental monitoring (Li et al., 2020; Ma et al., 2019).

Some of the major drivers are as follows:

1. **Data Overload:** The huge amount of geospatial data, such as satellite images, drone images, and ground sensor data, is far too huge to be analyzed by a human brain. AI/ML tools take care of managing and analyzing humongous amounts of big data to extract valuable insights (Ballantyne et al., 2021).
2. **Real-Time Analysis:** Disaster response and urban planning are some of the sectors that require real-time data analysis for quick decision-making. AI and ML can analyze data on the fly so that its insights are as timely as possible, which is critical for such applications (Nogueira et al., 2020).
3. **Precision and Accuracy:** It should continue to precision for discovering patterns, anomalies, and changes in geospatial data that can be useful for applications like land use classification, resource management, and environmental conservation. AI and ML algorithms thereby significantly improve the accuracy of such analyses (Zhu et al., 2017).

4. **Advances in Computing Power:** Recent advancements in hardware, cloud computing, and AI frameworks have enabled large-scale geospatial datasets to be processed at significantly greater speeds and with more efficiency. This opens up new avenues for the real-time interpretation of data and predictive analytics (Reichstein et al., 2019; Ghorbanzadeh et al., 2021).
5. **Economic and Environmental Impact:** Geo-spatial analysis can reduce the cost of developing urban infrastructure, optimized agriculture practices, and the proper management of natural resources. Economics and environmental impacts therefore lead to the use of AI and ML in this field (Li et al., 2020; Zhao et al., 2021).

The motivation for applying AI and ML in geospatial analysis would be to provide smarter, faster, and more accurate solutions that enable better decision-making and sustainable development practices across multiple industries.

Applications Driving Motivation:

The motivation for integrating AI and ML into geospatial analysis is further reinforced by their wide range of applications, including:

1. **Environmental Monitoring:** AI and ML are used to monitor changes in the environment, such as deforestation, urbanization, and climate change. This helps in making informed decisions to protect natural resources and mitigate environmental degradation (Cheng et al., 2020).
2. **Disaster Management:** AI and ML play a crucial role in disaster preparedness, response, and recovery. They enable rapid damage assessment, predictive modeling, and resource allocation during natural disasters like earthquakes, floods, and hurricanes (Khosravi et al., 2022).
3. **Urban Planning:** AI and ML assist in urban development by providing insights into land use patterns, infrastructure planning, and traffic management. This helps city planners make data-driven decisions to create sustainable and resilient urban environments (Batty, 2018).
4. **Agriculture:** AI and ML technologies are transforming agriculture by enabling precision farming, crop health monitoring, and yield prediction. This leads to more efficient use of resources and higher agricultural productivity (Kamilaris & Prenafeta-Boldú, 2018).

In summary, the motivation for integrating AI and ML into geospatial analysis stems from the need to process large volumes of data, improve accuracy and efficiency, and enable real-time decision-making. These technologies are driving innovation across various industries, leading to smarter, faster, and more accurate solutions for complex geospatial challenges.

6. OBJECTIVE

AI and ML application in geospatial data analytical work is in the creating of effective and smart systems that can fully automate the process of complex data interpretation for various areas of decision making adoption. Hence, these technologies will also facilitate management and analysis of extensive geospatial data as well as increasing the accuracy of forecasts and efficient gathering of relevant information for use (Yao et al., 2018; Zhang & Zhang, 2020).

The primary objective of integrating Artificial Intelligence (AI) and Machine Learning (ML) into geospatial data analysis is to create intelligent systems that can automate the interpretation of complex datasets, enabling more accurate and efficient decision-making across various domains. By leveraging AI and ML, researchers and practitioners aim to enhance the capabilities of geospatial analysis, making it more robust, scalable, and applicable to real-world problems (Cheng et al., 2017; Goodfellow et al., 2016).

Some specific objectives of AI/ML in geospatial analysis are:

1. **Design an Automated System for Interpretation of Data:** There is already a constructional element which is novel and free in the sense of good detection of patterns, land uses analysis and features of satellite images components localization while reducing the time used and eliminating the number of mistakes performed (Cheng et al., 2017; Ronneberger et al., 2015).
2. **Improved Prediction Models:** Making it leaner brings forth the use of Convolutional Neural Networks and Recurrent Neural Networks in advanced machine learning where the emphasis is on increasing the prediction of events within mainly natural calamities, urbanization, and in farming (He et al., 2016; Mnih et al., 2015).
3. **Integrate and Use Several Types of Data:** Create strategies on the use of all geospatial input, for instance, satellite images, and aerial images, in addition to any data collection devices such as ground-based sensors into a common analytical mode to enhance comprehension (Reichstein et al., 2019).
4. **AI/ML Models: Develop Scalable AI/ML Models:** Develop AI/ML systems that support various range of geospatial data from demanding and localized data to large and widespread satellite data (Ball et al., 2017).

5. **Sustainable Development Goals:** AI remains critical as well in the use of spatial data, within the context of improving resource utilization and management, assessing environmental changes and trends within and beyond borders, as well as predicting sustainability issues at the global level (Rolnick et al., 2019).

The main aim is to examine how AI and machine learning can revolutionize the field of geospatial data analytics by primarily making it more resourceful, precise, and competent in solving real-world problems.

Key Objectives in Detail

1. Automated Data Interpretation:
   - **Land Cover Classification:** Develop models that can automatically classify land cover types, such as forests, urban areas, and water bodies, from satellite imagery (Zhang & Zhang, 2020).
   - **Object Detection:** Create systems that can detect specific objects, such as buildings, roads, and vehicles, in geospatial images (He et al., 2016).
   - **Feature Extraction:** Automate the extraction of critical features from raw geospatial data, reducing the need for manual feature engineering (Ronneberger, Fischer, & Brox, 2015).
2. Enhanced Predictive Modelling:
   - **Disaster Prediction:** Use historical data to predict natural disasters, such as earthquakes, floods, and hurricanes, enabling better preparedness and response (Mnih, Kavukcuoglu, & Silver, 2015).
   - **Urban Growth Forecasting:** Develop models that can predict urban expansion and its impact on infrastructure and resources (Yao, Yang, & Lin, 2018).
   - **Agricultural Yield Prediction:** Use geospatial data to forecast crop yields, helping farmers optimize their practices and improve productivity (Zhang & Zhang, 2020).
3. Data Integration:
   - **Multi-Source Data Fusion:** Combine data from satellites, drones, and ground sensors to create a holistic view of the study area (Chalon & Spoon, 2007).
   - **Time-Series Analysis:** Integrate time-series data to monitor changes over time, such as deforestation, urbanization, and climate change (Zhang & Zhang, 2020).
4. Scalability and Adaptability:

- **Localized Analysis:** Develop models that can analyze small-scale, localized datasets for specific applications, such as precision agriculture or infrastructure monitoring (Goodfellow, Bengio, & Courville, 2016).
- **Global-Scale Analysis:** Create systems capable of processing large-scale satellite imagery for global applications, such as climate monitoring or disaster management (Yao et al., 2018; He et al., 2016).

5. Sustainable Development:
    - **Resource Management:** Use AI and ML to optimize the use of natural resources, such as water, soil, and forests, ensuring their sustainable management (Zhang & Zhang, 2020).
    - **Environmental Conservation:** Monitor environmental changes and predict their impact on ecosystems, enabling proactive conservation efforts (Ronneberger et al., 2015).
    - **Disaster Resilience:** Develop strategies for disaster management that enhance the resilience of communities and infrastructure (Mnih et al., 2015; Yao et al., 2018).

7. IMPLEMENTATION

When creating a system that incorporates AI and machine learning for geospatial data analysis, it is undertaken in a methodological way and a number of essential steps are taken to warrant a well functioning system. The implementation of AI and ML in geospatial data analysis involves a systematic approach to ensure the development of robust and efficient systems (Goodfellow, Bengio, & Courville, 2016). This process includes several key steps, from designing the development environment to deploying the final model. Below, we explore the implementation process in detail:

1. Design and Development Environment:

    Hardware and Software Setup: The first step in implementing an AI/ML-based geospatial analysis system is to set up the necessary hardware and software environment. This includes high-performance computing resources, such as GPUs, and software frameworks like TensorFlow, PyTorch, and Keras (He et al., 2016).

    Data Storage and Management: Efficient data storage solutions, such as cloud-based platforms or distributed file systems, are essential for handling large geospatial datasets. Tools like Google Earth Engine (GEE) and Amazon S3 can be used for storing and managing geospatial data (Zhang & Zhang, 2020).

2. Data Preparation:

   Data Collection: Gather geospatial data from various sources, such as satellites, drones, and ground-based sensors. Ensure that the data is of high quality and relevant to the analysis.

   Data Cleaning: Preprocess the data to remove noise, correct errors, and handle missing values. Techniques like noise reduction, binarization, and edge detection are commonly used in this step.

   Data Augmentation: Enhance the dataset by applying transformations such as rotation, flipping, and scaling. This increases the diversity of the training data and improves the model's generalization ability (Zhang & Zhang, 2020).

3. Model Training:

   Dataset Splitting: Divide the dataset into training, validation, and test sets. A typical split might be 70% for training, 15% for validation, and 15% for testing (Goodfellow, Bengio, & Courville, 2016).

   Model Selection: Choose the appropriate machine learning model based on the task. For image-based tasks, Convolutional Neural Networks (CNNs) are commonly used, while Recurrent Neural Networks (RNNs) are suitable for time-series data (Hochreiter & Schmidhuber, 1997).

   Training the Model: Train the model on the training dataset using frameworks like TensorFlow or PyTorch. Define the model architecture, compile it with an optimizer (e.g., Adam or SGD), and train it over multiple epochs (Abadi et al., 2016; Paszke et al., 2019).

   Hyperparameter Optimization: Tune hyperparameters such as learning rate, batch size, and number of epochs to optimize the model's performance (Bergstra & Bengio, 2012).

4. Model Evaluation:

   Performance Metrics: Evaluate the model's performance using metrics like accuracy, precision, recall, and F1 score. Confusion matrices can also be used to visualize the model's performance (Sokolova & Lapalme, 2009).

   Validation and Testing: Validate the model on the validation set and test it on the test set to ensure it generalizes well to unseen data (Goodfellow et al., 2016).

5. Model Deployment:

   Integration with GIS: Deploy the trained model by integrating it with Geographic Information Systems (GIS) or other geospatial platforms. This allows users to upload geospatial data and receive real-time analysis results (Zhou et al., 2020).

Real-Time Applications: Implement the model in real-time applications, such as disaster response systems or urban planning tools, where timely insights are critical (Yuan, Liu, & Zhang, 2021)

7.1 Design and Development Environment

Unless stated specifically, it's best to assume that the following components fall under the hardware and software design and development environment. These steps of implementation procedures are some of the important aspects that call for the effective formulation of a strong AI/ML based geospatial analytics system.

1. **Machine Learning Model:** For a certain task available different types of models can be used such as Random Forest, Support Vector Machines, Convolutional Neural Networks, etc. Every model is advantageous depending on the usage, for instance, Random Forest is used for classification tasks while CNN models are efficient on image-based tasks (Krizhevsky, Sutskever, & Hinton, 2012).
2. **Training Data Aspects:** The other technique deals with rotation, flipping, and scaling of the data set. These are meant for enhancing the variety of the training data being fed into the machine. This aids in improving the generalization ability of the model so as to reduce overfitting (Shorten & Khoshgoftaar, 2019).
3. **Assessment of the Construct:** This is a measure of how well the outcomes of a statistical analysis will be used beyond the data on which it was based. This includes training a model on a portion of the data and evaluating it on the remaining portion without training on this validation set so that the model parameters can be adjusted for optimal performance of the model (Goodfellow, Bengio, & Courville, 2016).

7.2 Data Preparation

The inclusion of AI/ML models with real-time applications also involves GIS and other data platforms. Dynamic updating of data, therefore, makes it relevant to real-time analysis and decision-making. For example, the integration of land cover classification model with GIS offers immediate information regarding the land use patterns and thus supports strategic urban planning decisions (Zhu et al., 2017).

Data preparation requires several operations to be performed on the geospatial datasets until it is clean, so that it can be ready to use. The most common set of operations includes the following:

- Image Processing Operations:

- **Noise Reduction:** The median filter is an operation that eliminates noise in an image without losing the edges (Gonzalez & Woods, 2018).
- **Grayscale Conversion:** Conversion of color images to grayscale reduces complications in recognition as it concentrates on light variations rather than color (Shapiro & Stockman, 2001).
- **Thresholding:** In this thresholding process, texts are separated from the background since the contrasting feature between characters and background is enhanced with grayscale images in the binary form (Otsu, 1979).
- **Line Segmentation:** Text is segmented into the projection methods used at lines for segmentation in horizontal direction (Pal & Chaudhuri, 2004).
- **Word Segmentation:** The projection algorithms have been applied vertically to find the spaces between words for segmentation (Pal & Chaudhuri, 2004).
- **Character Segmentation:** For simplicity, divide each word into characters and then analyze the model on individual pieces (Pal & Chaudhuri, 2004).
- **Image Segmentation:** The segmented images are subdivided into lines, words, and characters for further processing (Shi & Malik, 2000).

Figure 5. Segmentation process in geospatial analysis

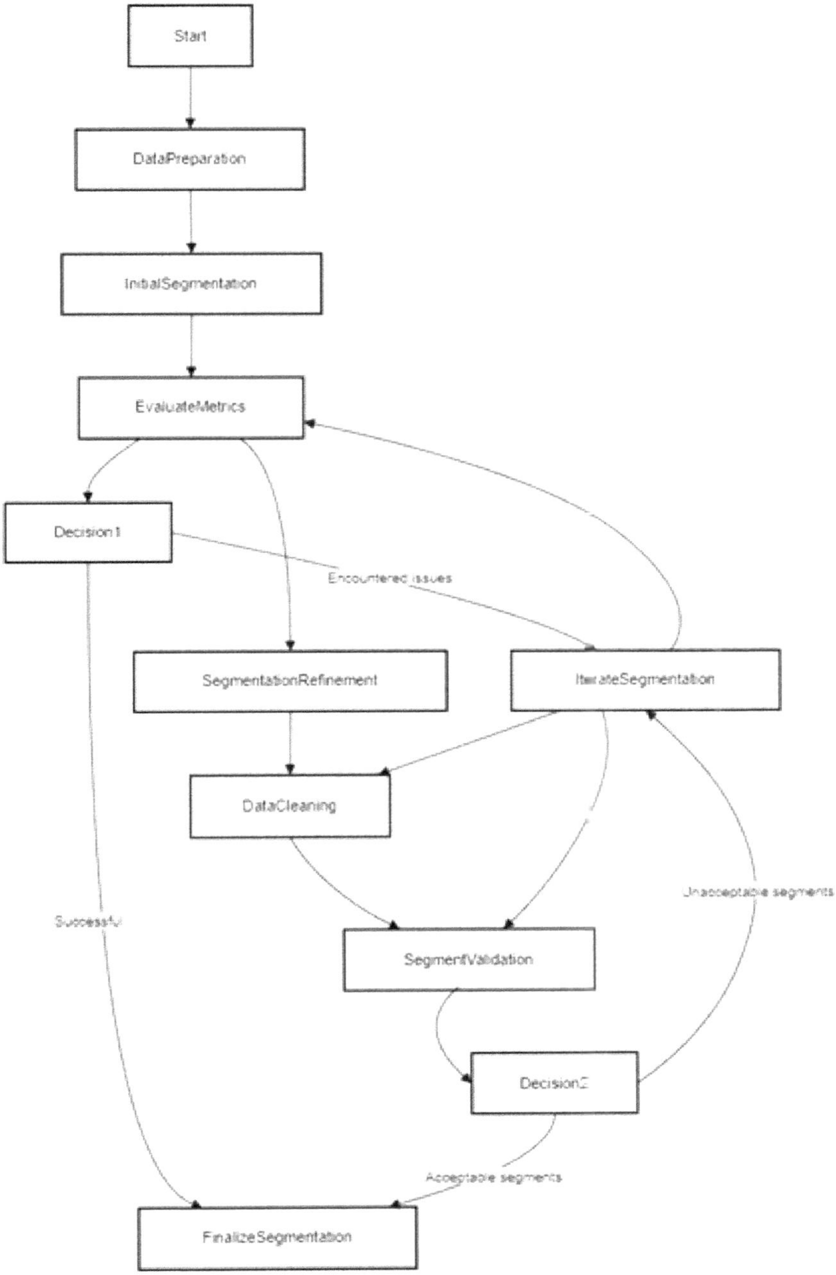

7.3 Model Training

Probably the most important implementation task for the model is training the model. The steps involved in this process are:

1. **Dataset Splitting:** Split the prepared dataset into training, validation and test sets. A usual split for the proportion is 70% to train, 15% for validation, and the remaining 15% for testing (Goodfellow et al., 2016).
2. **Training the Model:** Model training is used to train on the training data by using the support of pre-existing frameworks that enable machine learning such as Tensorflow or Pytorch. Which involves defining the structural requirement of the model, for example, the number of layers, types of layers e.t.c., then compiling the model using an optimizer like Adam or SGD and training the data on the training dataset (Abadi et al., 2016).
3. **Hyperparameter Optimization:** Here, hyperparameters refer to the treatment like learning rate, batch size, number of epochs, etc. All of these should be optimized in order to enhance the performance of the model (Bergstra & Bengio, 2012).
4. **Model Evaluation:** The fitted model is run on the validation set to test its performance, and performance metrics are computed like how accurate, precise, recall, and F1 score. Also, performance of the models can be represented through confusion matrices (Sammut & Webb, 2017).

Table 3. Evaluation metrics for model performance

Metric	Description
Accuracy	The ratio of correctly predicted instances to the total instances.
Precision	The ratio of true positive predictions to the total predicted positives.
Recall	The ratio of true positive predictions to the total actual positives.
F1 Score	The harmonic mean of precision and recall, providing a balance between the two.

7.4 Model Evaluation

Model evaluation is a critical step in the AI/ML lifecycle, ensuring that the trained model performs well on unseen data and meets the desired accuracy and reliability standards. After training the model, it is essential to evaluate its performance using a separate test dataset that was not used during training or validation. This step helps in assessing the model's generalization ability and identifying potential issues such as overfitting or underfitting. Key evaluation metrics for geospatial analysis tasks include accuracy, precision, recall, F1 score, and confusion matrices. For regression tasks, metrics like Mean Squared Error (MSE), Root Mean Squared Error (RMSE), and R-squared are commonly used. Additionally, visual evaluation techniques, such as overlaying model predictions on satellite imagery or maps, can provide intuitive insights into the model's performance (Zhang & Zhang, 2020). If the model does not meet the desired performance thresholds, further tuning of hyperparameters, data augmentation, or even revisiting the feature extraction process may be necessary. Ultimately, thorough model evaluation ensures that the AI/ML system is robust, reliable, and ready for deployment in real-world geospatial applications.

7.5 Model Deployment

Once the model is adequately trained and tested, it is ready for actual usage. Such system can be utilized for example by embedding it within some preexisting system or application allowing users to upload geospatial data for further processing. The implementation of AI and ML in geospatial data analysis involves a systematic approach, from setting up the development environment to deploying the final model. This process ensures the creation of robust and efficient systems that can handle complex geospatial datasets and provide valuable insights for decision-making (Goodfellow, Bengio, & Courville, 2016).

8. APPLICATION OF AI/ML IN GEOSPATIAL DATA

The integration of Artificial Intelligence (AI) and Machine Learning (ML) into geospatial data analysis has led to transformative applications across various domains. These technologies enable the automation of complex tasks, improve the accuracy of predictions, and provide actionable insights for decision-making. Below, we explore some of the key applications of AI and ML in geospatial data analysis:

8.1 Environmental Monitoring

Environmental monitoring is one of the most critical applications of AI and ML in geospatial analysis. The ability to process large volumes of satellite imagery and sensor data in real-time has revolutionized how we monitor and protect the environment.

- **Deforestation Monitoring:**

- AI models, particularly CNNs, are used to analyze satellite imagery to detect deforestation in real-time. For example, the Global Forest Watch platform uses AI to monitor forest cover changes and alert authorities to illegal logging activities (Cheng, Han, & Lu, 2017).
- These models can differentiate between natural forest loss (e.g., due to wildfires) and human-induced deforestation, enabling targeted conservation efforts.

- **Climate Change Analysis:**
  - AI algorithms analyze historical climate data to predict future climate scenarios. For instance, ML models can predict sea-level rise, temperature changes, and extreme weather events (Yao, Yang, & Lin, 2018).
  - Geospatial data is also used to monitor the melting of polar ice caps and glaciers, providing critical insights into the impacts of global warming.
- **Wildlife Conservation:**
  - AI-powered drones and satellite imagery are used to monitor wildlife habitats and track animal movements. For example, ML models can identify and count endangered species, such as elephants or tigers, from aerial images (Zhang & Zhang, 2020).
  - These technologies help conservationists protect biodiversity and combat poaching.

8.2 Disaster Management

AI and ML have become indispensable tools for disaster management, enabling faster and more accurate responses to natural disasters.

- **Earthquake Damage Assessment:**
  - After an earthquake, AI models analyze satellite imagery to assess the extent of damage to buildings, roads, and infrastructure. For example, the xView2 challenge by the Defense Innovation Unit (DIU) used AI to automate damage assessment from satellite images (Cheng et al., 2017).
  - These models can prioritize areas for rescue and recovery efforts, saving lives and resources.
- **Flood Prediction and Management:**
  - ML algorithms analyze historical flood data and real-time sensor data to predict flood-prone areas. For instance, Google's Flood Forecasting Initiative uses AI to provide early warnings to communities at risk (Yao et al., 2018).
  - During floods, AI models can analyze satellite imagery to map affected areas and guide relief efforts.

- **Wildfire Detection and Management:**
  - o AI models analyze satellite imagery and sensor data to detect wildfires in their early stages. For example, NASA's Fire Information for Resource Management System (FIRMS) uses AI to provide real-time wildfire alerts (Goodfellow et al., 2016)

These models can also predict the spread of wildfires based on weather conditions and terrain, helping firefighters plan containment strategies.

8.3 Urban Planning

AI and ML are transforming urban planning by providing data-driven insights into land use, infrastructure, and transportation.

- **Smart Cities:**
  - o AI models analyze geospatial data to optimize urban infrastructure, such as energy grids, water supply systems, and waste management. For example, AI can predict energy demand and optimize power distribution in real-time.
  - o Smart city initiatives, such as Singapore's Smart Nation project, use AI to improve urban living and sustainability (Zhang & Zhang, 2020).
- **Traffic Management:**
  - o AI algorithms analyze traffic patterns from satellite imagery and GPS data to optimize traffic flow and reduce congestion. For example, AI-powered traffic lights can adapt to real-time traffic conditions, reducing wait times and emissions.
  - o These models can also predict traffic bottlenecks and suggest alternative routes for commuters.
- **Land Use Planning:**
  - o AI models classify land cover types from satellite imagery, enabling urban planners to make informed decisions about zoning and development. For example, ML models can identify areas suitable for residential, commercial, or green spaces (He, Zhang, Ren, & Sun, 2016).
  - o These insights help create sustainable and resilient urban environments.

8.4 Agriculture

The AI/ML technologies alter the agricultural practices in the following activities as shown below:

- **Crop Health Monitoring**: AI models analyze satellite imagery to monitor crop health and detect diseases or pest infestations. This help farmers take timely action to protect their crops and improve yields.
- **Precision Agriculture**: Geospatial data is used to optimize the use of resources, such as water, fertilizers, and pesticides. AI systems provide recommendations for irrigation, planting, and harvesting, leading to more sustainable farming practices.
- **Yield Prediction**: Machine learning algorithms predict crop yields based on historical data and environmental conditions. This information helps farmers plan their operations and manage risks associated with climate variability (Zhang & Zhang, 2020).

8.5 Natural Resource Management

- **Water Resource Management**: AI models analyze geospatial data to monitor water bodies, assess water quality, and predict water availability. This information is used for managing water resources and ensuring sustainable usage.
- **Forest Management:** Geospatial data is used to monitor forest health, detect illegal logging, and plan reforestation efforts. AI systems provide insights into forest dynamics and support conservation initiatives.
- **Mineral Exploration**: AI and ML algorithms analyze geological data to identify potential sites for mineral exploration. This reduces the cost and environmental impact of mining activities (Ronneberger, Fischer, & Brox, 2015).

8.6. Public Health

- **Disease Surveillance:** Geospatial data is used to track the spread of infectious diseases and identify high-risk areas. AI models predict disease outbreaks and support public health interventions (Yao et al., 2018).
- **Healthcare Access:** AI systems analyze geospatial data to assess access to healthcare facilities and identify underserved areas. This information is used to plan the allocation of medical resources and improve healthcare delivery.

In summary, the applications of AI and ML in geospatial data analysis are vast and transformative. These technologies enable more accurate, efficient, and automated analysis of complex datasets, leading to better decision-making across various domains, including environmental monitoring, disaster management, urban planning, agriculture, and public health.

9. CHALLENGES IN AI/ML FOR GEOSPATIAL ANALYSIS

While AI and ML have revolutionized geospatial analysis, several challenges remain:

1. **Data Quality and Availability:**
    - Geospatial data often suffers from issues such as noise, missing values, and inconsistencies. Pre-processing this data to ensure quality is a time-consuming and resource-intensive task (Yao, Yang, & Lin, 2018).
    - Additionally, access to high-quality geospatial data can be limited, particularly in developing regions (Zhang & Zhang, 2020).
2. **Computational Complexity:**
    - Processing large-scale geospatial datasets requires significant computational resources. Training deep learning models on high-resolution satellite imagery can be computationally expensive and time-consuming (Goodfellow, Bengio, & Courville, 2016).
    - Edge computing and cloud-based solutions are being explored to address these challenges (Cheng, Han, & Lu, 2017).
3. **Model Interpretability:**
    - Many AI models, particularly deep learning models, are often considered "black boxes" due to their lack of interpretability. This is a significant challenge in geospatial analysis, where decisions can have far-reaching social and environmental impacts (Ronneberger, Fischer, & Brox, 2015).
    - Explainable AI (XAI) techniques are being developed to address this issue, but more work is needed to make these models transparent and trustworthy (Srivastava, Pant, & Agarwal, 2020).
4. **Ethical and Privacy Concerns:**
    - The use of AI in geospatial analysis raises ethical concerns, such as data privacy and bias. For example, satellite imagery can inadvertently capture sensitive information, such as private property or individual movements (Zheng, Zhang, Ma, Xie, & Ma, 2011).
    - Ensuring that AI models are fair and unbiased is critical, particularly in applications like urban planning and disaster management.
5. **Integration of Heterogeneous Data:**
    - Geospatial data often comes from diverse sources, such as satellites, drones, and ground sensors. Integrating this data into a unified analytical framework is a significant challenge (Yao et al., 2018).
    - Future research should focus on developing techniques for seamless data integration and fusion.

10. EMERGING TRENDS IN AI/ML FOR GEOSPATIAL ANALYSIS

1. **Edge Computing for Real-Time Analysis:**
   - Edge computing involves processing data closer to its source, such as on satellites or drones, rather than in centralized data centers. This approach reduces latency and enables real-time analysis, which is critical for applications like disaster response and environmental monitoring (He et al., 2016).
   - For example, AI models deployed on drones can analyze imagery in real-time to detect wildfires or assess flood damage.
2. **Explainable AI (XAI):**
   - Explainable AI techniques aim to make AI models more transparent and interpretable. In geospatial analysis, XAI can help users understand how models make decisions, such as classifying land cover or predicting disasters (Srivastava, Pant, & Agarwal, 2020).
   - Techniques like SHAP (SHapley Additive exPlanations) and LIME (Local Interpretable Model-agnostic Explanations) are being explored for geospatial applications.
3. **Generative AI for Data Augmentation:**
   - Generative AI models, such as Generative Adversarial Networks (GANs), can create synthetic geospatial data for training machine learning models. This is particularly useful in scenarios where real-world data is scarce or difficult to obtain (Goodfellow, Bengio, & Courville, 2016).
   - For example, GANs can generate synthetic satellite imagery to augment training datasets for land cover classification.
4. **AI-Driven Automation:**
   - AI is being used to automate various aspects of geospatial analysis, from data preprocessing to model deployment. For example, automated feature extraction tools can reduce the need for manual intervention, speeding up the analysis process (Cheng, Han, & Lu, 2017).
   - AI-driven automation is also being used to optimize workflows in GIS platforms, making geospatial analysis more accessible to non-experts.
5. **Integration with IoT and Smart Devices:**
   - The integration of geospatial data with IoT devices and smart sensors is enabling new applications in areas like precision agriculture and smart cities. For example, IoT sensors can provide real-time data on soil moisture, which can be combined with satellite imagery to optimize irrigation (Zheng et al., 2011).

- This integration is creating a more interconnected and data-rich environment for geospatial analysis.

11. FUTURE DIRECTIONS

The field of AI and ML in geospatial data analysis is rapidly evolving, with new advancements and applications emerging regularly. As these technologies continue to mature, several future directions hold promise for further enhancing the capabilities and impact of geospatial analysis. Below, we explore some of the key areas for future research and development:

1. **Hybrid Models:**
   - **Combining Traditional and Deep Learning Approaches**: Future research could focus on developing hybrid models that combine traditional machine learning techniques with deep learning approaches. These models could leverage the strengths of both methods, such as the interpretability of traditional models and the feature extraction capabilities of deep learning, to achieve more robust and accurate results (Zhang et al., 2020).
   - **Ensemble Methods:** Ensemble methods, which combine multiple models to improve performance, could be explored for geospatial analysis. For example, combining CNNs with Random Forests or Support Vector Machines (SVMs) could enhance the accuracy of land cover classification or object detection tasks (Cheng et al., 2017).
2. **Edge Computing in Geospatial Analysis:**
   - **Real-Time Processing:** Moving AI/ML processing closer to data sources, such as satellite sensors or UAVs, could enable real-time analysis and decision-making. Edge computing reduces latency by processing data locally, making it ideal for applications like disaster response and environmental monitoring (Gao et al., 2015).
   - **Resource Efficiency:** Edge computing can also reduce the computational load on central servers, making it more efficient for large-scale geospatial projects. This approach is particularly useful for remote areas with limited connectivity (Zheng et al., 2011).
3. **Explainable and Transparency in AI:**
   - **Transparency and Interpretability:** As AI models become more complex, there is a growing need for explainable AI (XAI) techniques that provide insights into how models make decisions. This is particularly

important for geospatial analysis, where decisions can have significant social, economic, and environmental impacts (Srivastava et al., 2020).

- **Ethical Considerations:** Explainable AI can help address ethical concerns related to bias, fairness, and accountability in geospatial analysis. By making AI models more transparent, researchers and practitioners can ensure that their applications are used responsibly (Rojas et al., 2023).

4. **Integration of Heterogeneous Data:**
   - **Multi-Source Data Fusion:** Future research should focus on integrating data from diverse sources, such as satellite imagery, social media, IoT devices, and ground-based sensors. This integration can provide a more comprehensive understanding of geospatial phenomena and improve the accuracy of predictions (Yao et al., 2018).
   - **Cross-Domain Applications:** Combining geospatial data with data from other domains, such as healthcare, transportation, and finance, could lead to innovative applications. For example, integrating geospatial data with healthcare data could improve disease surveillance and resource allocation (Li et al., 2013).
5. **Ethical and Social Implications:**
   - **Data Privacy:** As geospatial data becomes more widely used, ensuring data privacy and security will be crucial. Future research should explore techniques for anonymizing data and protecting sensitive information (Chalon & Spoon, 2007).
   - **Bias and Fairness:** AI models can inadvertently perpetuate biases present in the training data. Future work should focus on developing techniques to detect and mitigate bias in geospatial analysis, ensuring that AI applications are fair and equitable (Freeman, 1994).
6. **AI-Driven Automation:**
   - **Automated Preprocessing:** Future developments could focus on automating the preprocessing of geospatial data, such as noise reduction, image segmentation, and feature extraction. This would reduce the time and effort required for manual preprocessing and improve the efficiency of geospatial analysis (He et al., 2016).
   - **Automated Model Selection:** AI-driven systems could automate the selection of machine learning models based on the characteristics of the dataset and the specific task. This would streamline the development process and make AI/ML more accessible to non-experts.
7. **Scalability and Adaptability:**
   - **Global-Scale Analysis:** Future research should focus on developing AI/ML models that can handle global-scale geospatial datasets. This would

enable applications like global climate monitoring, disaster prediction, and resource management (Goodfellow et al., 2016).

- **Localized Applications:** At the same time, AI/ML models should be adaptable to localized datasets for specific applications, such as precision agriculture or urban planning. This would ensure that the models are relevant and effective across different geographic regions.

12. FINAL THOUGHTS AND SUMMARY

The integration of Artificial Intelligence (AI) and Machine Learning (ML) into geospatial data analysis has opened up new possibilities for understanding and managing the Earth's resources. These technologies have transformed the field, enabling more accurate, efficient, and automated analysis of complex datasets. As we conclude this comprehensive exploration of AI and ML in geospatial analysis, let's recap the key takeaways and reflect on the broader implications of these advancements.

1. **Revolutionizing Geospatial Analysis:**
    - AI and ML have revolutionized geospatial analysis by automating the interpretation of large and complex datasets. Tasks such as land cover classification, object detection, and image segmentation, which were once time-consuming and error-prone, can now be performed with high accuracy and efficiency (Zhang et al., 2020).
    - Deep learning models, particularly Convolutional Neural Networks (CNNs), have become indispensable tools for analyzing satellite imagery and other geospatial data. Their ability to automatically learn hierarchical features from images has significantly reduced the need for manual feature engineering (Cheng et al., 2017).
2. **Applications Across Diverse Domains:**
    - The applications of AI and ML in geospatial analysis span a wide range of domains, including environmental monitoring, disaster management, urban planning, agriculture, and public health (Srivastava et al., 2020). These technologies enable more informed decision-making and contribute to sustainable development.
    - For example, AI models can monitor deforestation, predict natural disasters, optimize urban infrastructure, and improve agricultural practices. These applications have a profound impact on society, economy, and the environment (Yao et al., 2018).
3. **Tools and Technologies:**

- A wide range of tools and technologies, such as TensorFlow, PyTorch, Google Earth Engine, and OpenCV, support the development and deployment of AI/ML models in geospatial analysis (Goodfellow et al., 2016). These tools provide a robust ecosystem for data processing, model training, and visualization.
- Python, with its extensive libraries like GDAL, Rasterio, and GeoPandas, has become the go-to programming language for geospatial data analysis (Ronneberger et al., 2015).

4. **Challenges and Future Directions:**
   - Despite the significant advancements, challenges remain in areas such as data integration, model interpretability, and ethical considerations. Future research should focus on developing hybrid models, edge computing solutions, and explainable AI techniques to address these challenges (Rojas et al., 2023).
   - The integration of heterogeneous data sources, such as satellite imagery, social media, and IoT devices, will further enhance the capabilities of geospatial analysis. Additionally, ensuring data privacy and addressing bias in AI models are critical for the ethical use of these technologies (Zheng et al., 2011).
5. **Impact on Sustainable Development:**
   - AI and ML have the potential to contribute significantly to sustainable development goals by optimizing resource utilization, monitoring environmental changes, and predicting sustainability challenges (Li et al., 2013). These technologies enable better management of natural resources, leading to more resilient and sustainable communities.
   - For example, precision agriculture powered by AI can reduce water and fertilizer usage, while AI-driven disaster management systems can improve response times and save lives (Gao et al., 2015).
6. **Empowering Researchers and Practitioners:**
   - The availability of powerful tools, frameworks, and resources has empowered researchers and practitioners to tackle complex geospatial challenges. Online courses, tutorials, and community forums provide valuable learning opportunities and foster collaboration within the geospatial and AI communities (Goodfellow et al., 2016).
   - By leveraging these resources, professionals can stay updated with the latest advancements and apply cutting-edge techniques to their work.

In conclusion, the integration of AI and ML into geospatial data analysis has transformed the field, enabling more accurate, efficient, and automated interpretation of complex datasets. These technologies have a wide range of applications, from environmental monitoring and disaster management to urban planning and agriculture. As AI and ML continue to evolve, they will play an increasingly important role in addressing global challenges and supporting sustainable development. By leveraging the tools, technologies, and resources discussed in this document, researchers and practitioners can unlock the full potential of geospatial data and contribute to a better understanding of our planet. The future of geospatial analysis is bright, and AI and ML will undoubtedly remain at the forefront of this exciting field

13. CONCLUSION

The application of artificial intelligence and machine learning in enhancing and improving overall geospatial data analysis is a high-value, high-usage area. As strides are made in the development of deep learning models from their existing state to more complex levels of models, there will be a heightened appreciation of the geospatial datasets and their relationships.

New technologies of neural networks and big data availability will also help facilitate better in providing solutions to problems like environmental monitoring, relief of disasters, and even urban farming – activities that require agricultural land. Future studies should be focused on problems of integration of data from heterogeneous sources, explicability and safety of AI/ML technologies in practical use. This technology will help us get a deeper understanding of our planet, so it can contribute to development and competent decisions (Liu et al., 2021).

The advancements in AI and ML, particularly deep learning models like Convolutional Neural Networks (CNNs), have significantly enhanced the capabilities of geospatial data analysis. These models can automatically learn hierarchical features from images, reducing the need for manual feature engineering and improving the accuracy of tasks such as land cover classification, object detection, and image segmentation. The availability of powerful frameworks like TensorFlow, PyTorch, and Google Earth Engine has further streamlined the development and deployment of AI/ML models in geospatial analysis (Goodfellow et al., 2016).

The key benefits of integrating AI and ML into geospatial analysis include:

1. **Automation:** AI and ML automate the interpretation of large and complex geospatial datasets, reducing the time and effort required for manual analysis. This enables faster decision-making and reduces the risk of human error.
2. **Accuracy and Precision:** AI and ML algorithms improve the accuracy and precision of geospatial analysis by detecting patterns and anomalies that are difficult to identify using traditional methods. This is particularly important for

applications like disaster management and environmental monitoring, where precision is critical.

3. **Scalability:** AI and ML models can handle large-scale geospatial datasets, making them suitable for global applications such as climate monitoring and disaster prediction. They can also be adapted to localized datasets for specific applications like precision agriculture.
4. **Real-Time Analysis:** AI and ML enable real-time analysis of geospatial data, providing timely insights for applications like disaster response and urban planning. This is crucial for making informed decisions in dynamic environments.
5. **Sustainable Development:** AI and ML contribute to sustainable development by optimizing resource utilization, monitoring environmental changes, and predicting sustainability challenges. These technologies support the creation of resilient and sustainable communities.

REFERENCES

Chalon, V. & Spoon, P. (2007). Trends in multi-document analysis.

Cheng, G., Han, J., & Lu, X. (2017). Remote Sensing Image Scene Classification by Deep Learning. *IEEE Transactions on Geoscience and Remote Sensing*, *55*(5), 2670–2681.

Cohen, E. & Foku, A. (1995). Classification of beautiful handwritten text IAP02.

Gonzalez, R., & Woods, R. (2008). *Digital Image Processing*. Pearson Prentice Hall.

Goodfellow, I., Bengio, Y., & Courville, A. (2016). *Deep Learning*. MIT Press.

He, K., Zhang, X., Ren, S., & Sun, J. (2016). Deep Residual Learning for Image Recognition.

Liu, Y., & Wang, Z. (2020). Machine Learning for Predictive Modeling of Geospatial Data: A Review. *Computers, Environment and Urban Systems*, *82*, 101489.

Liu, Y., & Xie, L. (2021). Machine Learning in Geospatial Data Processing: A Review of Algorithms and Applications. *Journal of Geographic Information Science*, *35*(2), 145–162.

Long, J., Shelhamer, E., & Darrell, T. (2015). Fully Convolutional Networks for Semantic Segmentation. *Proceedings of the IEEE Conference on Computer Vision and Pattern Recognition (CVPR),* 3431-3440.

Mnih, V., Kavukcuoglu, K., Silver, D., Rusu, A. A., Veness, J., Bellemare, M. G., Graves, A., Riedmiller, M., Fidjeland, A. K., Ostrovski, G., Petersen, S., Beattie, C., Sadik, A., Antonoglou, I., King, H., Kumaran, D., Wierstra, D., Legg, S., & Hassabis, D. (2015). Human-level control through deep reinforcement learning. *Nature*, *518*(7540), 529–533. DOI: 10.1038/nature14236 PMID: 25719670

Müller, A., & Jain, R. (2020). Enhancing Geospatial Data Integration with AI/ML Techniques. *International Journal of AI and Data Mining*, *9*(1), 22–39.

Nian, N., Suk, H. I., & Kim, H. J. (2013). HOG feature enhancement for segmented handwritten character recognition.

Ronneberger, O., Fischer, P., & Brox, T. (2015). U-Net: Convolutional Networks for Biomedical Image Segmentation.

Rosa, F., & Bernardino, A. (2019). Advances in Deep Learning for Satellite Image Classification. *Remote Sensing*, *11*(8), 973.

Schmitt, C., & Kwang, H. (2021). A Survey on Artificial Intelligence and Machine Learning in Geospatial Data Processing. *Geo-Spatial Information Science*, *24*(3), 115–131.

Shen, X., & Zhang, C. (2019). A Review on Convolutional Neural Networks for Remote Sensing Image Classification. *International Journal of Applied Earth Observation and Geoinformation*, *79*, 12–23.

Yang, J., & Wang, C. (2017). Deep Learning for Geospatial Image Analysis: Applications and Challenges. *Remote Sensing*, *9*(3), 281.

Yao, X., Yang, Y., & Lin, H. (2018). Geospatial Data Analytics Using Machine Learning Techniques: A Survey. *Geospatial Intelligence Review*, *17*(2), 90–112.

Zhang, H., & Li, W. (2018). An Overview of Geospatial Big Data and AI Technologies. *Geospatial Data Science*, *42*(4), 58–75.

Zhang, L., & Zhang, L. (2020). Deep Learning for Geospatial Data Analysis: A Review. *Geospatial Data Science*, *42*(3), 45–58.

Zhu, X., & Wang, H. (2020). AI for Geospatial Intelligence: Applications and Challenges. *Geospatial Intelligence Journal*, *27*(3), 231–245.

ADDITIONAL READINGS

Google Earth Engine Documentation. (2025). Developers and About Page. https://developers.google.com/earth-engine

Open, C. V. Documentation. (2025). Documents. https://docs.opencv.org

PyTorch. (2025). Tutorials. https://pytorch.org/tutorials

TensorFlow. (2025). Tutorials. https://www.tensorflow.org/tutorials

QGIS. (2025). Resources Page. https://docs.qgis.org/

APPENDIX I

Example Code for CNN Implementation

To provide a practical example of how AI and ML can be applied to geospatial data analysis, we include a simple implementation of a Convolutional Neural Network (CNN) using TensorFlow and Keras. This example demonstrates how to build, train, and evaluate a CNN model for geospatial image classification.

Below is a Python code snippet that demonstrates how to implement a CNN for geospatial data classification. This example assumes that you have a dataset of geospatial images (e.g., satellite imagery) and corresponding labels (e.g., land cover types).

```
```

Import necessary libraries

import tensorflow as tf

from tensorflow.keras import layers, models

from tensorflow.keras.preprocessing.image import ImageData-
Generator

from sklearn.model_selection import train_test_split

import numpy as np

Define the CNN model architecture

def create_cnn_model(input_shape, num_classes):
```
```

```
model = models.Sequential()

# First convolutional layer

model.add(layers.Conv2D(32, (3, 3), activation='relu', in-
put_shape=input_shape))

model.add(layers.MaxPooling2D((2, 2)))

# Second convolutional layer

model.add(layers.Conv2D(64, (3, 3), activation='relu'))

model.add(layers.MaxPooling2D((2, 2)))

# Third convolutional layer

model.add(layers.Conv2D(128, (3, 3), activation='relu'))

model.add(layers.MaxPooling2D((2, 2)))

# Flatten the output and feed it into fully connected layers

model.add(layers.Flatten())

model.add(layers.Dense(128, activation='relu'))
```

```
    model.add(layers.Dense(num_classes, activation='softmax')) # Assuming 'num_classes' for classification

    return model

    # Load and preprocess the dataset

    # Replace this with your actual dataset loading code

    # X = ... # Geospatial images (e.g., satellite imagery)

    # y = ... # Corresponding labels (e.g., land cover types)

    # X_train, X_test, y_train, y_test = train_test_split(X, y, test_size=0.2, random_state=42)

    # Normalize the image data (if not already normalized)

    # X_train = X_train / 255.0

    # X_test = X_test / 255.0

    # Define the input shape and number of classes

    input_shape = (128, 128, 3) # Example: 128x128 RGB images

    num_classes = 10 # Example: 10 land cover classes
```

```
# Create the CNN model

model = create_cnn_model(input_shape, num_classes)

# Compile the model

model.compile(optimizer='adam',

loss='sparse_categorical_crossentropy', # Use 'categorical_
crossentropy' if y is one- hot encoded

metrics=['accuracy'])

# Train the model

# Replace this with your actual training data

# history = model.fit(X_train, y_train, epochs=10, valida-
tion_data=(X_test, y_test))

# Example of data augmentation (optional)

datagen = ImageDataGenerator(

rotation_range=20,

width_shift_range=0.2,
```

```
height_shift_range=0.2,

horizontal_flip=True,

vertical_flip=True

)

# Fit the model using data augmentation

# history = model.fit(datagen.flow(X_train, y_train, batch_size=32), epochs=10, validation_data=(X_test, y_test))

# Evaluate the model

# test_loss, test_acc = model.evaluate(X_test, y_test)

# print(f"Test Accuracy: {test_acc}")

# Save the model for future use

# model.save('geospatial_cnn_model.h5')
```

Explanation of the Code

1. **Model Architecture:** The CNN model consists of three convolutional layers followed by max-pooling layers. These layers extract hierarchical features from the input images. The output of the convolutional layers is flattened and fed into fully connected (dense) layers, which perform the final classification.
2. **Data Preprocessing:** The input images are normalized (scaled to the range [0, 1]) to improve model training. Data augmentation techniques, such as rotation, shifting, and flipping, are applied to increase the diversity of the training data and reduce overfitting.
3. **Model Training:** The model is compiled using the Adam optimizer and sparse categorical cross-entropy loss (for integer labels) or categorical cross-entropy loss (for one-hot encoded labels). The model is trained on the training dataset and validated on the test dataset.
4. **Model Evaluation:** After training, the model's performance is evaluated on the test dataset using metrics like accuracy.
5. **Model Saving:** The trained model is saved to a file (`geospatial_cnn_model.h5`) for future use or deployment.

Applications of the CNN Model

This CNN model can be applied to various geospatial tasks, such as:

- **Land Cover Classification:** Classifying satellite images into different land cover types (e.g., forest, urban, water).
- **Object Detection:** Detecting specific objects, such as buildings or roads, in geospatial images.
- **Disaster Damage Assessment:** Assessing the extent of damage after natural disasters using satellite imagery.

In summary, this example demonstrates how to implement a CNN for geospatial data analysis using TensorFlow and Keras. By adapting this code to your specific dataset and task, you can build powerful AI models for a wide range of geospatial applications.

APPENDIX II

Recommended Tools and Resources

To further support your work in AI and ML for geospatial data analysis, we provide a list of recommended tools, libraries, and resources. These tools are widely used in the geospatial and machine learning communities and can help streamline your workflow, from data preprocessing to model deployment.

1. Python Libraries for Geospatial Analysis:
   - GDAL/OGR: A powerful library for reading and writing geospatial data formats. It supports a wide range of raster and vector data formats. [GDAL Documentation](https://gdal.org)
   - Rasterio: A library for reading and writing geospatial raster data. It is built on top of GDAL and provides a more Pythonic interface. [Rasterio Documentation](https://rasterio.readthedocs.io)
   - GeoPandas: An extension of Pandas for working with geospatial data. It simplifies the manipulation of vector data, such as shapefiles. [GeoPandas Documentation](https://geopandas.org)
   - Shapely: A library for manipulation and analysis of planar geometric objects. It is often used in conjunction with GeoPandas. [Shapely Documentation](https://shapely.readthedocs.io)
2. Machine Learning Frameworks:
   - TensorFlow: An open-source machine learning framework developed by Google. It is widely used for building and training deep learning models. [TensorFlow Documentation](https://www.tensorflow.org)
   - PyTorch: An open-source machine learning framework developed by Facebook. It is known for its flexibility and dynamic computation graph. [PyTorch Documentation](https://pytorch.org)
   - Keras: A high-level neural networks API, written in Python and capable of running on top of TensorFlow. It simplifies the process of building and training deep learning models. [Keras Documentation](https://keras.io)
3. Geospatial Platforms:
   - Google Earth Engine (GEE): A cloud-based platform for planetary-scale geospatial analysis. It provides access to a vast repository of satellite imagery and geospatial datasets.

[Google Earth Engine Documentation](https://developers.google.com/earth-engine)

- QGIS: An open-source Geographic Information System (GIS) that supports the visualization, analysis, and editing of geospatial data.
  [QGIS Documentation](https://docs.qgis.org)

4. Data Visualization Tools:
   - Matplotlib: A plotting library for Python that provides a wide range of static, animated, and interactive visualizations.
     [Matplotlib Documentation](https://matplotlib.org)
   - Plotly: A graphing library that makes interactive, publication-quality graphs online. [Plotly Documentation](https://plotly.com/python)
   - Folium: A Python library for creating interactive maps using Leaflet.js. It is particularly useful for visualizing geospatial data.
     [Folium Documentation](https://python-visualization.github.io/folium)
5. Data Augmentation and Preprocessing:
   - OpenCV: An open-source computer vision library that provides tools for image processing, filtering, and segmentation.
     [OpenCV Documentation](https://docs.opencv.org)
   - Albumentations: A library for image augmentation that is optimized for performance and supports a wide range of transformations.
     [Albumentations Documentation](https://albumentations.ai)
6. Cloud Computing Platforms:
   - Google Cloud Platform (GCP): A suite of cloud computing services that includes tools for data storage, machine learning, and geospatial analysis.
     [Google Cloud Documentation](https://cloud.google.com/docs)
   - Amazon Web Services (AWS): A comprehensive cloud computing platform that offers services for data storage, machine learning, and geospatial analysis.
     [AWS Documentation](https://aws.amazon.com/documentation)
   - Microsoft Azure: A cloud computing platform that provides a wide range of services for data storage, machine learning, and geospatial analysis.
     [Azure Documentation](https://docs.microsoft.com/en-us/azure)
7. Online Courses and Tutorials:
   - Coursera: Offers courses on geospatial analysis, machine learning, and AI. Some recommended courses include:
     [Geospatial Analysis with Python] (https://www.coursera.org)

[Deep Learning Specialization by Andrew Ng] (https://www.coursera.org/specializations/deep-learning)

- o edX: Provides courses on geospatial data analysis and machine learning.
  [Geospatial Data Analysis with Python] (https://www.edx.org)
- o Kaggle: A platform for data science competitions and tutorials. It offers hands-on tutorials for machine learning and geospatial analysis.
  [Kaggle Tutorials](https://www.kaggle.com/learn)

8. Community and Forums:
   - o Stack Overflow: A Q&A platform for programmers. It is a valuable resource for troubleshooting and learning from the community.
     [Stack Overflow](https://stackoverflow.com)
   - o GIS Stack Exchange: A Q&A platform specifically for GIS professionals and enthusiasts.
     [GIS Stack Exchange](https://gis.stackexchange.com)
   - o Reddit: Communities like r/MachineLearning and r/gis are great for discussions and sharing knowledge.
     [r/MachineLearning](https://www.reddit.com/r/MachineLearning)
     [r/gis](https://www.reddit.com/r/gis)

Chapter 4
Approaches in Smart Urban Planning With Artificial Intelligence Through Data Science

Gunseerat Kaur
https://orcid.org/0000-0003-1577-9566
Lovely Professional University, India

ABSTRACT

Sustainability is a key aspect of preserving for future generations without hindering the needs of the current generation. Sustainability is of interest because it is regarded as a social goal. Ever since the migration of population into urban settings has shown increase it is a key aspect for sustainability to plan establish and run smart cities and implement sustainability goals while planning for further growth. Recent trends in artificial intelligence have led to a proliferation of studies that involve data and molding data for algorithms for better solutions. Data science and Artificial intelligence can create, control, and predict trends, solutions and issues involving urban planning, sustainability, and smart cities. This chapter discusses the use cases, problems and sustainable solutions and modes of using artificial intelligence and data science in urban planning and smart infrastructure.

INTRODUCTION

Artificial intelligence (AI) is the technology which has assisted in multitude of research by creating solutions and predictions to help humankind; Data Science (DS) is a key field which is working towards creation of relevant information and datasets

DOI: 10.4018/979-8-3373-3206-2.ch004

which help AI in understanding the problem statement and create solutions (Rathore et al., 2020). Together these technologies throw light on various scenarios that can be analysed for an issue, determining causes, effects, benefits of an approach. The idea is to inculcate sustainable development goals by researching ideas and implementations with AI and DS. Gradual evolution of technology and its involvement in building infrastructure can be observed currently with the introduction of smart homes, smart traffic systems, smart vehicles, or perhaps smart cities (Satpathy et al., 2025).

The main ideology of introducing these key terms in urban planning is to infuse technology advancements with functional infrastructure, making it more functional with less resources. With the onset of initial digital city creation for Amsterdam in 1994 (Boopathi, 2024), the growth gradually introduced key points of analysing urban data and management of cities. Traffic control systems were initial systems that were updated to manage increasing traffic and suggestions of shorter routes (Almuhanna & Salman, 2018). Implicitly this laid groundwork for other systems like parking systems. The development of smart infrastructure has complex relativity with multiple factors, such as cost, innovation, policies, and technology (Köster et al., 2018). To understand the importance of urban planning in this chapter explores the AI-driven approaches fine-tuned with data science to revolutionize the urban infrastructure and architecture (Hussain, 2024). Figure 1 discusses the brief technological advancements that shape the map of collecting data through sensors, analysing them with big data techniques and securing them with multiple strategies that can lead to usage of this data for deriving favourable results. The available data through satellite imagery, social media, sensors, IoT has proven to be a catalyst in figuring out solutions to make urban settlements better (Almolhis et al., 2020; Vangala et al., 2020). Planning cities that are more functional and resilient towards factors like transportation, housing, and availability of services. Algorithms in deep learning provide wider insights in data collected from multiple aspects, promotes result driven approach. Smart traffic systems are currently implemented in multiple countries which have shown reduction in traffic congestion, efficacy in alternative routes for fuel savings, smarter parking access and efficient lane and speed controls(Di Febbraro et al., 2020). Similarly, smart transportation systems are also braving better routes, decreasing latency in freight deliveries.

Figure 1. Technical components of the chronology of smart infrastructure

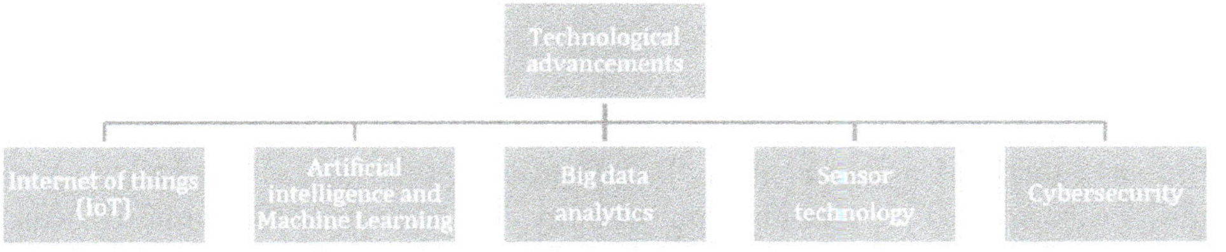

(Alahakoon et al., 2023; Rathore et al., 2020; Xiong et al., 2024)

The technological advancements bring innovations like Internet of Things (IoT) and sensor technology that provide real-time information and allow monitoring of factors. The data produced through the sensors and devices can be further analysed by algorithms created through machine learning (ML) and trends can be studied to predict with big data analytics (Kamrowska-Załuska & Bibri, 2021). These sophisticated technologies provide effective projects that transform the notion of understanding smart infrastructure roadmap better. The smart infrastructure was innovated from various strategies such as laws, society and advancements in techniques and tools. The gradual upliftment of the economy also impacted the growth of smart features. Urban planning and Smart infrastructure require decision making and case-based analysis of previous information (Korada, 2021). Figure 2 emphasizes different factors responsible for growth of Smart infrastructure. Mainly it involves AI tools and their usage, which can be utilized in techniques like survey, navigation, prediction, obstacle avoidance, data collection and data processing, monitoring, etc. (Ranyal et al., 2022). It allows us to make informed and proof-based decisions which provide a basis for collection and analysis of data. Recently, AI has been applicable to revolutionize almost all industries and sectors, this is also evident in planning for urban architectures and methods for reducing urban infrastructure challenges. Multiple metropolitan cities across the world are using smart systems to enhance their traffic challenges. Evidence from (Javaid et al., 2018), (Sharma et al., 2020) show effective traffic management can provide much needed solutions. Smarter cities need to identify and predict the dynamic situations that is easily managed by machine learning algorithms and data collected from multiple sources. The capacity to recognise similar patterns and similar situations in large datasets can be managed with data science strategies (Gkontzis et al., 2024), allowing for key breakthroughs and changing the methods of planning for future usage. To further enhance the functionality, these smart systems can also aid towards achieving the sustainability goals for future usage. Convolutional neural networks (CNNs) are effective in searching greener spaces and water bodies (Butt et al., 2023), making it optimal to ploy the needs of housing and industry placements in infrastructure planning. (Kourtit, 2021) employed clustering techniques to monitor and predict water distribution techniques to reduce water loss in city's water grid.

It is critical to understand that rapid urbanization has led to shift the paradigm of rural dwellers transitioning towards urban setups. The carbon emissions and resource consumption has increased and this has led to unplanned architectures or poorly schemed infrastructures. The disadvantage is increasing challenges of transportation, food production, employability, and imbalanced lifestyles (Estrada & Vargas, 1 C.E.; Mehmood & Ning, 2024). To foster equality and accessibility it is necessary to understand the advantages of plotting the smart cities well to create affordable housing, reducing traffic congestion to preserve ecological balance and sustainability code.

Figure 2. Factors influencing the growth of smart infrastructure

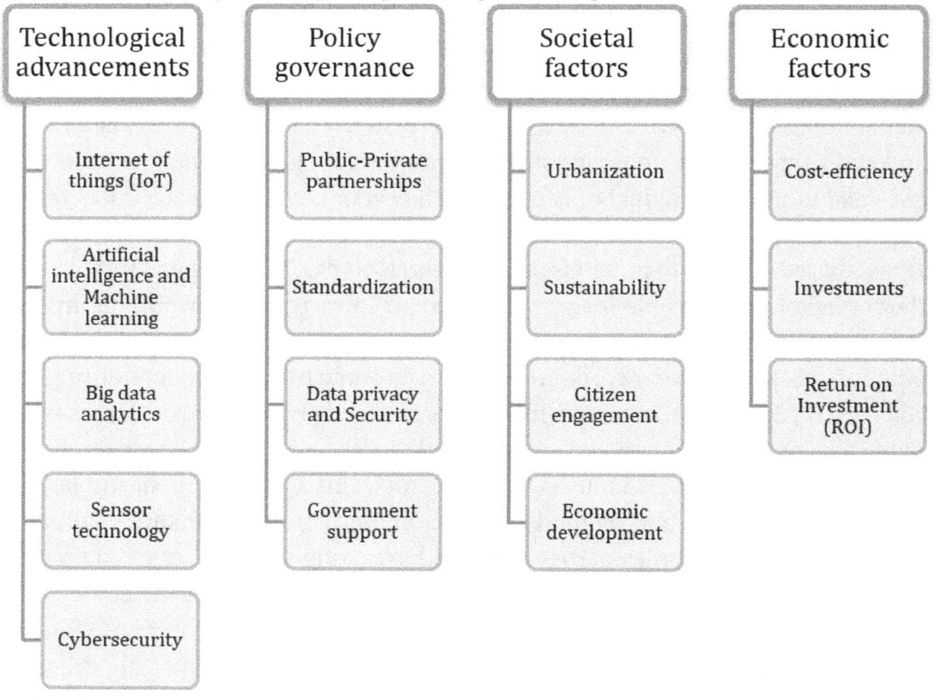

(Estrada & Vargas, 1 C.E.; Golubchikov & Thornbush, 2020; Yao et al., 2019)

This data is moreover used for providing information for natural language processing which further enhances the understanding of smart infrastructure towards humans and cater better solutions. Gathering the major ideologies of creating smart infrastructure it is important to focus on systems of transportation, which allows for ease of access in spaces of habitat (Rizwan et al., 2017). Evidently, cities have incorporated digitalization in traffic analysis, population analysis and other urban planning matters from earlier times; However, rising networking and computerization of control systems has produced far better opportunities to explore (Javaid et al., 2018). The focus has now shifted to creating more automation control systems and better urban planned infrastructure systems. Considering sustainable urban planning majorly four factors are focussed upon with aims to create communities with environmentally and socially equitable and sustainable (Mahmoud et al., 2022). Importance is towards dependable urban transportation, green infrastructure, usage of renewable energy, use of methods resilient to climate change need to be combined in city design and management to create liveable and self-sustaining communities (Hashem et al., 2023). Figure 3 enlists the key factors affecting sustainable urban planning.

Figure 3. Planning factors for sustainable urban planning

Environmental sustainability	Economic sustainability	Social sustainability
• Green infrastructure • Climate change adaptation • Waste management • Renewable energy • Air and Water quality	• Affordable housing • Job creation • Public-Private partnerships	• Walkable and bikeable communities • Public transportation • Community engagement • Public health and safety

(Abbass et al., 1 C.E.; Hashem et al., 2023; Mehmood & Ning, 2024)

Promoting renewable energy and green infrastructure will create alternatives to fossil fuels, furthermore it is a favourable step in the climate crisis. Using AI, Cities like Singapore (*LTA | Intelligent Transport Systems*, n.d.) and New York (*NYC DOT - Infrastructure - Traffic Signals*, n.d.) are fostering community centric planning to analyse traffic patterns and commutes. The smart traffic management systems work towards proper lane monitoring and real-time traffic management. This provides better fuel usage and reduces jams across busy cities. (Grishina et al., 2021) have used AI-driven systems to collect and route waste to proper recycling facilities. Similar usage can predict better results and avoid hassle for city dwellers. This brings about better resource efficiency and promotes a circular economy. By planning cities well, it will eventually catalyse economic growth, accelerate businesses, stimulate regional developments, thereby creating job market and raising economic standards (Askari et al., 2019).

This chapter aims to research the multiple methods, case studies and relevance of using AI and DS in smart urban planning. It is divided into eight parts with introduction to factors and their significance in building the ideology of urban planning through sustainable ways, second part will state the research question that the chapter will answer while progressing. Parts three onwards till six will emphasize on topics with evidence-based research, discussing the previous methodologies and research carried out. The seventh part of the chapter will focus on challenges and concerns, finally the last part will conclude.

RESEARCH OBJECTIVES

The objective of this chapter is to bridge the gap between advance AI methods and smart systems and current urban dwelling infrastructures. The identified practical approaches have been used in different sectors to generate results in smaller niche, the smart systems are used in traffic management, planning housing, solving water crisis, managing plantations and environment. The chapter focuses on answering three major research questions;

RQ1: What are the major factors in Smart Urban Planning that can be enhanced with AI & DS?
RQ2: What methods are mainly used on data that creates results and methods used by AI for enhancing the Smart Urban infrastructure?
RQ3: What challenges and relevant mitigation strategies exist in planning for smarter infrastructure?

While synthesizing through vast research work of multiple authors, these questions were formulated to understand how effective these approaches can be. The discussion will emphasize on technical as well as sustainability efforts too, keeping in account the ethical, political, and social implications.

DEFINING SMART CITY: CORE OBJECTIVE OF TRANSITION FROM TRADITIONAL PLANNING

Understanding the concept of smart city is very lucrative theoretically, however, following and fulfilling each concepts takes a properly planned approach and vast learning. A smart city can be easily profiled as an urban dwelling region that has been carefully crafted while addressing the issues of current urban cities, by improving quality of life of individuals(Khediri et al., 1 C.E.), by improving fuel efficiency, creating job markets, stabilizing economy, and proving sustainability goals. At core, a smart city is a combination of technology-based decisions in favour of environment and its dwellers(Rathore et al., 2020). Clearly, this is not the case in traditional cities, where the criteria of choosing or creating a service was based on either the political or social influence. Traditional approach was more based on historic information or relying on previously implemented plans that were potentially less rational. Non-dynamic nature of traditional planning loses in front of data-driven approaches in smart urban planning (Kamrowska-Załuska & Bibri, 2021), also the goals differ for each criterion. Table 1 discusses multiple factors and differences in traditional and smart urban planning approach. This comparison helps us understand the actual factors that need AI and DS for enhancing urban planning, answering our first research question to begin with.

Table 1. Traditional v/s smart urban planning approach

Feature	Traditional Urban Planning	Smart Urban Planning
Focus	Primarily on physical development and infrastructure	Integrates ICT across city services and infrastructure
Data Usage	Relies on historical data, demographic projections, manual surveys	Heavily relies on real-time data from sensors, connected devices, digital platforms
Technology Integration	Limited or no integration of advanced digital technologies	Deeply integrates IoT, AI, big data, cloud computing, mobile applications
Efficiency	Can be less efficient in resource management and service delivery	Aims to maximize efficiency in resource consumption and service delivery
Connectivity	May not prioritize or fully leverage digital connectivity	Prioritizes robust digital connectivity and communication networks
Stakeholder Involvement	Citizen engagement might be less emphasized	Emphasizes the involvement of diverse stakeholders through digital platforms
Adaptability	Can be less adaptable to rapid changes	Designed to be more adaptable and responsive to changing conditions
Sustainability Focus	Sustainability might be a consideration	Places a strong emphasis on environmental sustainability through smart technologies
Cost	Lower upfront costs, potentially higher long-term operational costs	Higher initial investments in technology, aims for lower long-term costs

(Gkontzis et al., 2024; Mortaheb & Jankowski, 2023; Rathore et al., 2020)

OVERVIEW OF AI AND DS IN TRANSFORMING URBAN SETTLEMENTS

Data science has been proved as a catalyst in dynamic sectors of technological innovation, Urban systems produce data that needs to be analysed, stored and monitored regularly. The factors, interacting variables, compiling results from gathered data are some of the relevant activities that are carried out upon receiving data (Rathore et al., 2020). This data can be collected through sources such as satellite imagery, sensors placed at multiple points or even edge devices and cloud-based devices. Unlike traditional planning methods, smart planning systems can rely on collected data and its patterns to recognise the need and predicted factors (Hosseinian-Far et al., 2017). Researchers have used (Tran et al., 2017) time-series and predictive algorithms to reduce traffic congestion in cities like Los Angeles (Rizwan et al., 2017). Such decision-making benefits overall environment by reducing carbon emissions, thereby creating smoother urban mobility. Another concept used is digital twin technology is an innovation used for depicting current urban settlements with merged reality of virtual settlement model. These models are created by simulating, predicting, planning, and mapping real-time data as per demographic statistics. This allows to simulate current city's challenges via a visual representation (Yassine et al., 2019), creating solutions to mitigate against said challenges. The integration of such digital models into urban planning has dramatically improved the reliability of predictive analytics and policy efficacy (Kourtit, 2021). Considering such examples, Figure 4 provides multiple aspects based upon which the smart urban developments can be planned.

Figure 4. Multifaceted Smart urban planning aspects

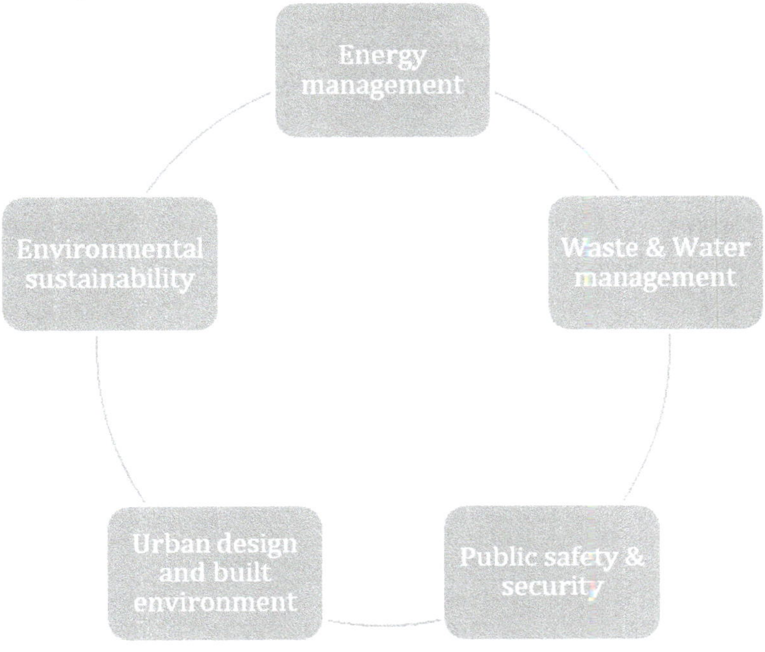

(Chang et al., 2020; Golubchikov & Thornbush, 2020)

ROLE OF ARTIFICIAL INTELLIGENCE AND DATA SCIENCE IN SMART URBAN PLANNING

To understand the methodology used by multiple researchers in vast instances, it is important to understand the importance of data science. Data science thorough techniques like data analytics creates answers from present data. The data is mainly collected through sensors, actuators, smart devices, edge nodes, satellite imaging(Mortaheb & Jankowski, 2023). The data from these sources is crucial for monitoring systems and create insights for future refences. The key part of the process is to provide data integration, as multiple sensors, energy meters, traffic sensors, public systems, come from different manufacturers and work on different techniques(Razavi et al., 2019). It is important to not loose data and perform exploratory data analysis beforehand to determine the factors data is governed by(Chang et al., 2020). Further, decision support systems are deployed to analyse current conditions and map factors which make or break decision making. After application of preprocessing pipelines and standardizing data, identification of data is performed to drop any unimportant data(S. Li et al., 2024). The advanced analytics create forecasting and prediction for data sought. The insights are created from final data after processing, creating visuals that help understanding the causes and effects(Yao et al., 2019). Combining these techniques the details are discussed in Table 1.

Table 2. Key techniques, roles in data science for urban planning

Technique/Approach	Role in Processing & Analysis	Example Applications
IoT Data Integration & Sensor Fusion(Almalki et al., 2021; Razavi et al., 2019)	Aggregates, cleans, and consolidates data from diverse urban sensors and devices to create a unified dataset for analysis.	Real-time traffic monitoring, energy consumption tracking, environmental sensing
Machine Learning & Predictive Analytics(Korada, 2021; Sanchez et al., 2024)	Applies statistical models and AI algorithms to derive forecasts, optimize resource allocation, and detect emerging trends from historical and real-time data.	Traffic congestion prediction, energy demand forecasting, predictive maintenance
Simulation & Optimization Models(Z. Li et al., 2019)	Uses computational models to simulate urban scenarios, test various infrastructural setups, and optimize layouts for efficiency and sustainability.	Urban growth planning, crisis response simulations, optimal public transport routing
Data Visualization & Decision Support Tools(Adiono et al., 2020; S. Li et al., 2024)	Translates complex processed data into visual dashboards and GIS maps to support informed decision-making by urban planners and city managers.	Smart city command centres, interactive urban planning dashboards

Considering artificial intelligence in urban planning, it supports goal-oriented decision making. AI algorithms can use insights created through data science to transform decisions dynamically (Feizi et al., 2020). AI powered models help in optimization of models, creating powerful monitoring and decision-making systems that allow complex tasks to be easily handled, such as energy grid handling, or traffic management (J. Li et al., 2022).

APPROACHES IN AI & DS

To answer second research question, it is pivotal to understand the methods and techniques deployed on raw data to create relevant results out of it. Smart traffic management systems are deployed initially to control the signals for traffic lights, after successful accuracy rates. Furthermore, it is used for managing parking assistance, alternative route finding and congestion avoidance. Long short-term memory (LSTM) models are effective in considering time-series analytics which are helpful in clustering, classification algorithms (Z. Li et al., 2019; Sharma et al., 2020). For more precise results, researchers have used deep learning algorithms on data, gaining more precision on dynamic data and showing better traffic management (Alghamdi, 2024). Machine learning models like decision trees, K-means neural networks have shown promising results in wireless sensor network-based city management with IoT based devices (Bandur et al., 2019; Hao et al., 2024). The researchers have used these techniques in pollution and traffic management for smart cities. Logistics and route management has been another forefront that is supporting smart industries, techniques like AI and IoT are combined with clustering methods in machine learning to perform route optimization and regulate traffic flow (Zhou et al., 2023). Researchers have applied a spectrum of techniques in combination with each other to gain better results. Computer vision is another crucial technique when the data collected requires dynamic video feed analysis. Dynamically, the results are more in real-time and provide spontaneous and relevant analysis when needed.

Figure 5. Techniques & methods for smart urban planning

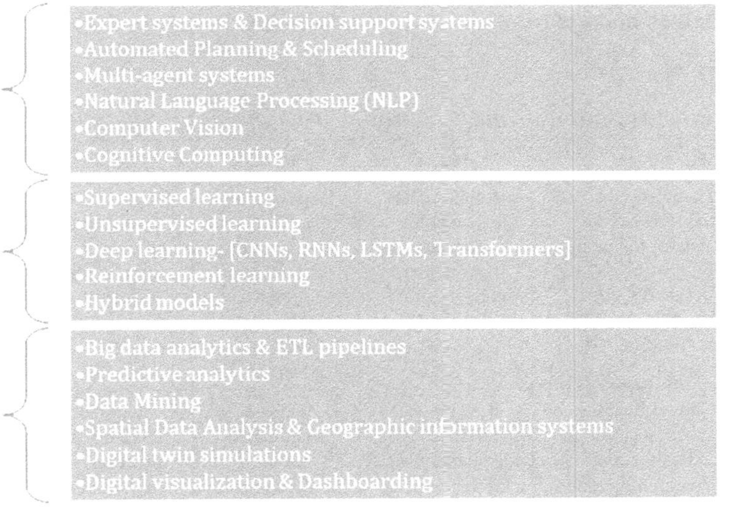

(Rabby et al., 2019; Yao et al., 2019)

This method is also effective in gauging assistance in detection of anomalies and monitor major changes through image-based datasets. Natural language processing (NLP) is used for gaining insights in text-based data, a case study by (Alahakoon et al., 2023) shows processing feedbacks received from city residents to improve the systems and make better decisions. NLP also aides in feeding this information to LLMs (Wang et al., 2021) to gain more awareness on sentiments. Figure 5 shows the methods that are used by researchers to work with smart urban planning. Considering reinforcement learning and gaining perception from cloud-based systems also provides optimized route optimizations improving rush hour traffic. Rule-based systems provide heuristics for waste management, segregating and recycling waste collected at multiple points in a city(Ren et al., 2021). User-aware smart home systems provide better efficient solutions for managing energy levels and usage in housing sector. (Boopathi, 2024) have shown usage of graphic neural networks (GNNs)(Yu et al., 2024) to reduce traffic congestions and efficient resource allocation for optimization of traffic management systems and parking assistance systems. As evident from table 3, A hybrid approach of combining blockchain with data science has improved traffic management accuracy by improving operational systems. (Majeed et al., 2014) have used cellular wireless techniques in reducing carbon emissions with ANNs and Naïve bayes strategies (Filho et al., 2020). Geographical information systems (GIS)(Yilmaz et al., 2014) provide pivotal information in spatial data analysis and shows significant results in digital twin simulations. Reflecting upon research work from previous years it is evident that smart mechanisms have provisioned much needed precision in decision making and accuracy in understanding factors needed to plan the urban dwellings.

Table 3. Findings and techniques from previous research work

Paper	Method	Findings	Results	Techniques Used
Demystifying the Role of Natural Language Processing (NLP) in Smart City Applications (Tyagi & Bhushan, 2023)	Applies NLP to analyse unstructured urban text (policies, public feedback)	Demonstrates that NLP can extract salient themes and sentiments to support data-driven urban governance	Improved processing efficiency and timely trend detection in urban documents	Text mining, sentiment analysis, topic modelling, Natural Language Understanding (NLU)
Natural Language Processing in Urban Planning: A Research Agenda (Fu, 2024)	Reviews and synthesizes NLP applications in urban planning literature	Finds a fragmented research landscape with high potential for standardized techniques and benchmarking	Outlines future research directions and opportunities to integrate NLP into urban analytics	Topic modelling (LDA), document clustering, sentiment analysis, text mining
Role of Computer Vision in Smart Cities: Applications and Research Challenges (Mehmood & Ning, 2024)	Surveys computer vision applications in urban environments (e.g., surveillance, infrastructure monitoring)	Validates computer vision's role in detecting anomalies and automating urban monitoring tasks	Demonstrates enhanced decision support through automated image and video analysis	Convolutional Neural Networks (CNNs), image segmentation, object detection, video analytics

continued on following page

Table 3. Continued

Paper	Method	Findings	Results	Techniques Used
Computer Vision for Creating Sustainable Smart Cities of Tomorrow (Hussain, 2024)	Examines the integration of computer vision frameworks for public service automation	Shows that leveraging CV improves public safety, traffic management, and spatial data accuracy	Reduced manual intervention and real-time monitoring improvements in urban settings	Deep learning, transformers, object tracking, anomaly detection
Smart City Urban Planning using an Evolutionary Deep Learning Model (Alghamdi, 2024)	Combines evolutionary optimization with deep recurrent neural networks for urban prediction	Successfully predicts urban growth patterns and optimal spatial layouts	Outperforms conventional heuristic methods in forecasting urban evolution and resource allocation	Recurrent Neural Networks (RNNs), Teaching–Learning-Based Optimization (TLBO), evolutionary algorithms
Artificial Intelligence Based Methods for Smart and Sustainable Urban Planning: A Systematic Survey (Koumetio Tekouabou et al., 2022)	Systematically reviews AI methods addressing urban planning problems	Identifies key themes across land use, mobility, urban growth, and sustainability challenges	Highlights the rapid adoption of ML and ensemble techniques in urban planning decision support	Machine learning algorithms, deep neural networks, ensemble learning
Machine Learning for Smart Cities: A Comprehensive Review of Applications and Opportunities (Son et al., 2023)	Reviews various machine learning applications in smart city contexts	Concludes that ML significantly enhances resource management and urban service delivery	Points to future challenges regarding scalability and interpretability issues	Supervised/unsupervised learning, deep learning, predictive modelling
Deep Reinforcement Learning for Traffic Flow Optimization in Urban Planning (Deshmukh et al., 2023)(Tan et al., 2022)	Develops a deep Q-learning framework to optimize urban traffic signals	Identifies notable reductions in traffic congestion and improved lane utilization	Achieves up to a 50% reduction in waiting times and 10% overall improvement in flow efficiency	Deep Q-Network (DQN), reinforcement learning, simulation modeling
AI Agent as Urban Planner: Steering Stakeholder Dynamics via Consensus-based Multi-Agent Reinforcement Learning (Qian et al., 2023)	Proposes a multi-agent reinforcement learning system for participatory planning	Shows that dynamic consensus among diverse stakeholders improves planning outcomes	Enhanced stakeholder satisfaction with more balanced and adaptive land use decisions	Multi-agent reinforcement learning, consensus mechanisms, graph neural networks

continued on following page

Table 3. Continued

Paper	Method	Findings	Results	Techniques Used
Spatial Planning of Urban Communities via Deep Reinforcement Learning (Zheng et al., 2023)	Frames spatial planning as a sequential decision-making problem on graph structures solved by RL	Generates urban layouts that outperform traditional human-designed plans in efficiency	Demonstrates superior planning outcomes under varying urban scenarios	Deep reinforcement learning, graph neural networks, simulation-based training
Enhancing Urban Resilience: Smart City Data Analyses, Forecasts, and Digital Twin Techniques (Gkontzis et al., 2024)	Integrates predictive analytics with digital twin simulations at the neighbourhood level	Reveals that real-time monitoring and simulation significantly aid resilience planning	Provides accurate forecasts and supports effective scenario-based interventions	Predictive analytics, digital twin technology, ETL processes, GeoDataFrame visualization
Innovative Approaches to Smart Cities and Urban Planning: A Case-Based Review of Global Best Practices and Challenges (Esfandi et al., 2024)	Conducts comparative case studies of smart city initiatives worldwide	Finds that centralized automation combined with inclusive, citizen-centric models yields optimal outcomes	Identifies challenges (e.g., scalability, digital equity) and recommends best practices	Case study analysis, comparative evaluation, qualitative review
Unlocking Urban Futures: The Role of Big Data Analytics and AI In Urban Planning – A Systematic Literature Review and Bibliometric Insight (Korada, 2021)	Uses a systematic literature review and bibliometric analysis of AI in urban planning	Establishes AI's capacity to forecast urban growth and optimize infrastructure resource allocation	Maps trend and reveals ethical considerations and collaboration patterns in the field	Bibliometric analysis, big data analytics, systematic review, AI algorithms
The Research Landscape of AI in Urban Planning: A Topic Analysis with ChatGPT (Sanchez et al., 2024)	Applies topic modelling combined with ChatGPT to analyse urban planning literature	Identifies 16 key research themes, from urban design to predictive modelling in transportation	Offers a comprehensive overview of current trends and research gaps in AI-enabled urban planning	Latent Dirichlet Allocation (LDA), NLP, AI-driven topic modelling

continued on following page

Table 3. Continued

Paper	Method	Findings	Results	Techniques Used
Urban Computing & Intelligent Decision Support for Sustainable Cities (Hashem et al., 2023)	Integrates large-scale urban data collection with decision support systems	Demonstrates that intelligent data fusion supports proactive urban management and resilience	Leads to improved strategic planning and real-time resource reallocation	Urban computing, simulation models, decision support systems, data analytics
AI-Driven Urban Planning Platforms: Transforming Cities into Smart, Sustainable Havens (Skubis et al., 2024)	Develops an integrated AI platform leveraging sensor data and predictive models for planning	Enhances forecasting accuracy and streamlines the deployment of urban services	Achieves better alignment between IT strategies and urban development milestones	AI forecasting, machine learning, real-time sensor analytics, data integration

CASE STUDIES

As already established from the above data, it is evident that AI and data is used for improving the urban infrastructure, it is important to consider the current implementations in real life urban cities. To promote more greener solutions cities like Copenhagen, Amsterdam from Europe have developed separate infrastructure to support cycling practices across the cities. Following similar architectures cities from Asia like Nanning have implemented similar solutions for promoting green practices by residents(Yu et al., 2024). Facing rapid growth in infrastructure, alters the city construction areas, subdividing into different areas, multiple cities across the world such as Singapore, Stockholm, Vancouver or Chandigarh have prioritized green cover within urban spaces for promoting ecological balance(Wolniak & Stecuła, 2024). Smart city like Zurich has presented better infrastructure and connectivity to its residents providing strong optical fibre infrastructure for connectivity and creating smart city aspect.

Here are some of the impactful models developed over the years:

Urban city from scratch: Songdo, South Korea presents a city that has been constructed from scratch with a simple aim to build sustainable smart city with efficient connectivity(PROJECT OVERVIEW – IBD, n.d.). Taking a technology, driven approach public and private organizations were brought together to leverage a strong connected area for providing business hub with sustainable and potential planned framework for office, commercial and residential activities. The results showed reduction in energy consumption due to proper positioning of green spaces and reduction of carbon footprint due to pedestrian and cycle friendly design for city. The business district approved only LEED-certified (Leadership in Energy and Environmental Design) design for construction, therefore improving on energy efficiency. Another AI based example is from treating local waste water in South Korea using AI, it supports analytical data for determining the chemicals required to treat current waste water also sorting any other wastes within water with the help of AI.

Model for collaborative governance: Amsterdam, Netherlands has remarkably shown commitment towards staying sustainable regardless of challenges. The city provides extensive incentives on adoption of renewable energy sources(Policy: Sustainability and Energy - City of Amsterdam, n.d.). The cycle lanes promote strong culture of using the bikes to reach within city. The city authorities have three agendas for its residents and private service providers in the city; the transition to sustainable energy, increasing green and water filled spaces in the city and lastly promoting circular transition to reduce amount of new raw materials creation, preventing depletion of resources.

Ecological balance and water needs: Singapore has shown the remarkable growth in applying technology in its city-wide environment. The city faces water scarcity to deal with which they have used Data science-based model, on an average day the city uses around 430 million gallons of water(How GovTech Uses AI to Enhance Digital Public Services, n.d.). The city uses smart water allocation system to determine areas of usage to allocate water with constraints. Linear programming model optimizes the parameters of water collection and filtration for provisioning back to residents(How Singapore Is Using AI | CIO, n.d.).

Smart traffic system implementations: Singapore's Land Transport Authority uses massive real-time monitoring systems to manage traffic across the 160 KM network of roadways(How Singapore Is Using AI | CIO, n.d.), the system utilizes data from sensors; to manage traffic flow and public transportation facilities, their connected and unified approach collects data from consumer apps, sensors and its operations control centre utilizes this information to develop planning and policies dynamically. Cities like Riyadh and Jeddah in Saudi Arabia are also applying similar strategies with AI-based monitoring to control bottleneck traffic reducing traffic hassles by 20%(Saudi Cities Go Smart: AI-Driven Traffic Management, n.d.). London, England has shown its patented AI based system for their traffic management. It features traffic control with AI-based data in real-time making it easier to dissolve bottle neck formations during rush hour, reducing commute with alternative route suggestions and generation of traffic signals as per any new plan in optimal time. It also addresses the issues of congestions by monitoring user-based data and assists in rolling emergency vehicles with forecasting(Simplifai AI Traffic Management - Digital Marketplace, n.d.).

Smarter systems and approaches: Zurich has implemented smart light systems to cut down energy consumptions in street lights which dim as per the flow of traffic, providing energy saving of up to 70%(AI & Cities: Risks, Applications and Governance | UN-Habitat, n.d.), these systems also act as connected WiFi connectivity points making it strongly connected city. Seoul, South Korea has worked towards helping their elder population by detecting any potential non-activity by monitoring the residents and alerting emergency services in case of need. They are also creating similar AI-detection mechanism to curb crime rates in the city.

CHALLENGES AND ETHICAL CONSIDERATIONS IN USING AI WITH DATA SCIENCE METHODS IN URBAN PLANNING FOR INFRASTRUCTURE

The potential of incorporating techniques with AI is seamless it has enabled researchers to create solutions which are effective and optimized. The improvements further pose several ethical challenges. The acquisition of large-scale data to train AI methods and creating predictions poses issues such as data privacy and legitimacy of sources, including interoperability and accountability issues (Bhatia & Ahanger, 2021). The concerns further extend towards limit of critique and quality of judging decision-making skills. Four major concerns identified are highlighted in figure 6, as data needs to be collected from various sources and origins in the city planned infrastructure; therefore, it is important to respect individual privacy and concerns upon surveillance. The pervasive nature of data collection that will be required to optimally run such huge smart systems will also need to be resilient against any threats (Shyamala Devi et al., 2019).

Figure 6. Challenges and ethical considerations in using AI

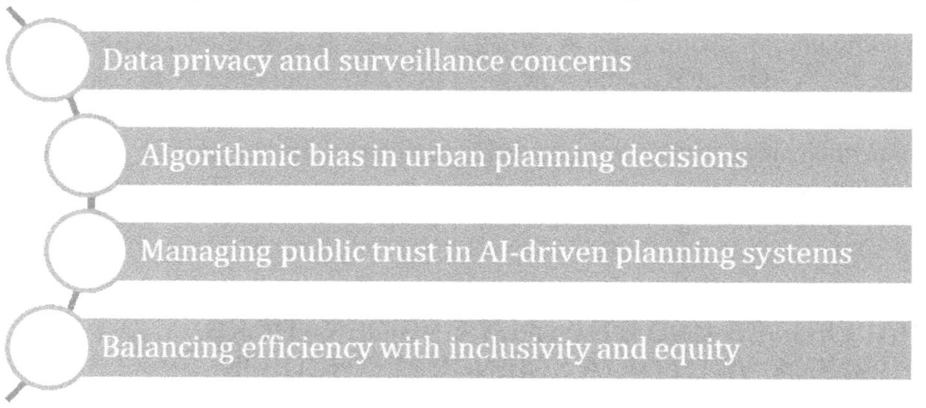

(Ali & Awad, 2018; J. Chen et al., 2021; Shyamala Devi et al., 2019)

Stringent practices need to be followed while accountability and legal ethics are concerned. Another issues that data science faces are bias in planning decisions (Kamrowska-Załuska & Bibri, 2021), as the nature of decisions is dynamic it is important for the data models to be free of bias. This is crucial for keeping inequalities in decision-making at bay. Often researchers have found that biased datasets have mis-predicted property valuations or mis-judged lane transitions in traffic management systems (Wallace et al., 2019). Mitigation in such scenarios require manual intervention and continuous monitoring. Significant barrier in implementing these vast city-wide technologies will be human mistrust and scepticism. The lack of transparency and accountability towards decision-making leads to public losing confidence in the system (Yassine et al., 2019). Initiatives must be taken to educate the users and urban dwellers to understand the new systems to promote adaptability instead of concerns. As the stakeholders in success of these urban smart cities and the regulations, it is wiser to involve public in decision-making and usage (Y. Y. Chen et al., 2019). Accounting for the potential hazards that are caused by AI are multiple (Yu et al., 2024), the barriers include people who happen to have trust issues, privacy and data security issues. Another aspect is high cost of hardware to be borne initially and then maintenance costs in subsequent years. Figure 7 discusses all the different paradigms of issues that mainly inhibit from wide adaptation of these techniques. Concerning about data privacy is mainly because data for accuracy needs to be collected at large in every form and this extensive data can lead to misuse or breach of data. Multi-dimensional systems which are interconnected to provide services, often tend to have different vulnerabilities, which if left unpatched can lead to data loss. The functioning and decision-making involved in AI systems can be difficult for masses, hence it causes concerns towards accountability that goes in hands of decision makers. Secondly, another concern put forward towards AI is the limitations in algorithms, the bias that can occur while training, this could lead to mistrust in real-time systems concerning people where physical safety concerns elevate.

Figure 7. Paradigms that AI adaptation slow

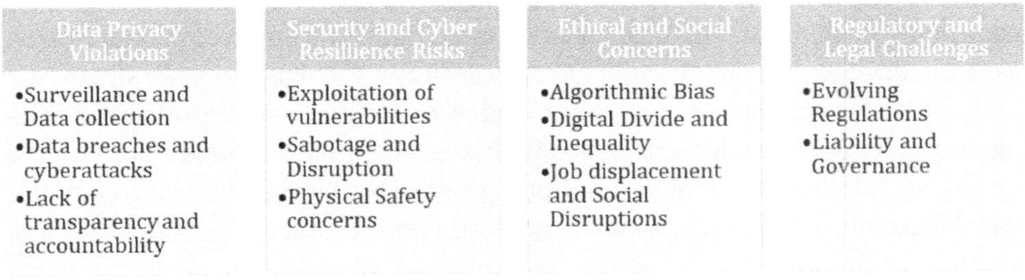

(Esfandi et al., 2024; Hashem et al., 2023; Skubis et al., 2024; Wolniak & Stecuła, 2024)

For the smart technologies to be adapted it is important to understand and teach everyone to understand the system well. (Wolniak & Stecuła, 2024) have elaborated six smart factors for smart cities; smart governance to allow proper handling of residents and creating balance between public-private organizations within city. Smart economy focussing on technology-based solution to leverage better performance and efficacy. Smart mobility and environment for residents of the city along with proper rule following and jurisdictions for accountability which will contribute to smart living. Smart people who understand the risks and benefits from technology implications and still make judicious use of these technologies (Qian et al., 2023). While adapting it is important to not let social disruptions between marginalized communities. The legal and regulatory bodies must be more careful with distributing responsibilities and accounting errors in judgement that can maximise further (Mahmoud et al., 2022). Creating balance without over-emphasizing technology is a pivotal point to achieve in making the ideas a success. To harness benefits, it is important to manage the ethical concerns over the decision-making processes.

CONCLUSION

As the rapid urbanization continues, it is evident that traditional approach of planned cities is reaching its potential and challenges are increasing in metropolitans. With the advent of AI, it is important to embrace the smarter methods and inculcate them into planning better urban dwelling cities which can mitigate solutions for current problems in terms of traffic, housing (Tao et al., 2018), environment, economy, and social front (Verma et al., 2018). The static approach of traditional cities is facing challenges on almost every front, ultimately challenging our approach towards attaining sustainability. Thereby a reason for exploring solutions through Artificial intelligence and Data science (Pharmacogn et al., 2018), which has proven significantly capable in mitigating solutions through hybrid approach. Current need is to design urban cities with proper planning that is dynamic and resilient, along being sustainable. This chapter aimed towards answering three major research questions through previous research. The results were combined to gather significant information for supporting answers as evidences for the research questions. The digital solutions gained through AI and DS can easily provide solutions for multiple aspects while planning urban cities. Some researchers have combined techniques like blockchain and IoT for enhancing the results and accuracy. The comprehensive approach in this chapter is exploring the foundational concepts and findings that have resulted in multiple approaches being incorporated to make better decisions. Table 4 comprehends the crux of previous discussion towards usage of AI and DS in urban planning.

Table 4. Summary of techniques

Technique/Approach	Results/Findings
Urban Computing & Intelligent Decision Support(Keerthana et al., 2021)	Combines big data collection, simulation models, and AI to form decision support systems. This integration enhances strategic and short-term planning, leading to more efficient resource allocation, improved sustainability performance, and increased urban resilience.
AI-Driven Urban Planning Platforms(Atta et al., 2020)	Leverages data from IoT devices, remote sensors, and social media along with machine learning algorithms to generate real-time insights. Results include enhanced forecasting of urban growth, optimized traffic management, and better public service deployment as seen in cities like Barcelona and Singapore.
Simulation & Optimization Models(Apthorpe et al., 2017)	Uses simulation algorithms and optimization techniques to model urban expansion and resource demands. These models help predict optimal urban layouts, reduce turnaround times in decision making, and generate sustainable urban design solutions.
Real-Time IoT Sensor Data Analytics(Seo et al., 2022)	Deploys networks of IoT sensors for continuous monitoring of key urban parameters (e.g., traffic, energy consumption, environmental metrics). This real-time data analysis facilitates dynamic management and quick responses to urban events, creating a more adaptive city environment.

REFERENCES

Abbas, M., Akhai, S., Abbas, U., Jafri, R., & Arif, S. M. (2025). Ai-enabled sustainable urban planning and management. In *Real-World Applications of AI Innovation* (pp. 233–260). IGI Global Scientific Publishing.

Adiono, T., Muttaqin, R., Rahayu, R. M., & Putra, R. V. W. (2020). *Cloud System Design using AMQP Protocol for Smart Devices System Applications*. DOI: 10.31227/osf.io/z824j

Alahakoon, D., Nawaratne, R., Xu, Y., De Silva, D., Sivarajah, U., & Gupta, B. (2023). Self-Building Artificial Intelligence and Machine Learning to Empower Big Data Analytics in Smart Cities. *Information Systems Frontiers*, *25*(1), 221–240. DOI: 10.1007/s10796-020-10056-x

Alghamdi, M. (2024). Smart city urban planning using an evolutionary deep learning model. *Soft Computing*, *28*(1), 447–459. DOI: 10.1007/s00500-023-08219-4

Ali, B., & Awad, A. I. (2018). Cyber and Physical Security Vulnerability Assessment for IoT-Based Smart Homes. *Sensors 2018, Vol. 18, Page 817, 18*(3), 817. DOI: 10.3390/s18030817

Almalki, F. A., Alsamhi, S. H., Sahal, R., Hassan, J., Hawbani, A., Rajput, N. S., Saif, A., Morgan, J., & Breslin, J. (2021). Green IoT for Eco-Friendly and Sustainable Smart Cities: Future Directions and Opportunities. *Mobile Networks and Applications 2021 28:1, 28*(1), 178–202. DOI: 10.1007/s11036-021-01790-w

Almolhis, N., Alashjaee, A. M., Duraibi, S., Alqahtani, F., & Moussa, A. N. (2020). The Security Issues in IoT-Cloud: A Review. *Proceedings - 2020 16th IEEE International Colloquium on Signal Processing and Its Applications, CSPA 2020*, 191–196. DOI: 10.1109/CSPA48992.2020.9068693

Almuhanna, A. (2018). Smart Traffic Management System for Traffic Control using Automated Mechanical and Electronic Devices. *IOP Conference Series. Materials Science and Engineering*, *377*(1), 012201. DOI: 10.1088/1757-899X/377/1/012201

Apthorpe, N., Reisman, D., Sundaresan, S., Narayanan, A., & Feamster, N. (2017). *Spying on the Smart Home: Privacy Attacks and Defenses on Encrypted IoT Traffic*. https://arxiv.org/abs/1708.05044v1

Askari, H., Khajepour, A., Khamesee, M. B., & Wang, Z. L. (2019). Embedded self-powered sensing systems for smart vehicles and intelligent transportation. *Nano Energy*, *66*, 104103. DOI: 10.1016/j.nanoen.2019.104103

Atta, A., Abbas, S., Khan, M. A., Ahmed, G., & Farooq, U. (2020). An adaptive approach: Smart traffic congestion control system. *Journal of King Saud University. Computer and Information Sciences*, *32*(9), 1012–1019. DOI: 10.1016/j.jksuci.2018.10.011

Banđur, Đ., Jakšić, B., Banđur, M., & Jović, S. (2019). An analysis of energy efficiency in Wireless Sensor Networks (WSNs) applied in smart agriculture. *Computers and Electronics in Agriculture*, *156*, 500–507. DOI: 10.1016/j.compag.2018.12.016

Bhatia, M., & Ahanger, T. A. (2021). Intelligent decision-making in Smart Food Industry: Quality perspective. *Pervasive and Mobile Computing*, *72*, 101304. Advance online publication. DOI: 10.1016/j.pmcj.2020.101304

Boopathi, S. (2024). Advancements in Machine Learning and AI for Intelligent Systems in Drone Applications for Smart City Developments. Https://Services.Igi-Global.Com/Resolvedoi/Resolve.Aspx?Doi=10.4018/978-1-6684-9596-4.Ch002, 15–45. DOI: 10.4018/978-1-6684-9596-4.ch002

Butt, H. A., Ahad, A., Wasim, M., Shayea, I., Coelho, P. J., Pires, I. M., & Garcia, N. M. (2023). Federated Machine Learning In 5G Smart Healthcare: A Security Perspective Review. *Procedia Computer Science*, *224*, 580–586. DOI: 10.1016/j.procs.2023.09.084

Chang, V., Sharma, S., & Li, C. S. (2020). Smart cities in the 21st century. *Technological Forecasting and Social Change*, *153*, 119447. DOI: 10.1016/j.techfore.2018.09.002

Chen, J., Ramanathan, L., & Alazab, M. (2021). Holistic big data integrated artificial intelligent modeling to improve privacy and security in data management of smart cities. *Microprocessors and Microsystems*, *81*, 103722. DOI: 10.1016/j.micpro.2020.103722

Chen, Y. Y., Lin, Y. H., Kung, C. C., Chung, M. H., & Yen, I. H. (2019). Design and implementation of cloud analytics-assisted smart power meters considering advanced artificial intelligence as edge analytics in demand-side management for smart homes. *Sensors (Basel)*, *19*(9), 2047. DOI: 10.3390/s19092047 PMID: 31052502

Di Febbraro, A., Gallo, F., Giglio, D., & Sacco, N. (2020). Traffic management system for smart road networks reserved for self-driving cars. *IET Intelligent Transport Systems*, *14*(9), 1013–1024. DOI: 10.1049/iet-its.2019.0675

Esfandi, S., Tayebi, S., Byrne, J., Taminiau, J., Giyahchi, G., & Alavi, S. A. (2024). Smart Cities and Urban Energy Planning: An Advanced Review of Promises and Challenges. *Smart Cities 2024, Vol. 7, Pages 414-444, 7*(1), 414–444. DOI: 10.3390/smartcities7010016

Estrada, E., & Vargas, M. P. M. (2021). Data Science Process for Smart Cities. In Latin American Women and Research Contributions to the IT Field (pp. 348-370). IGI Global Scientific Publishing.

Feizi, A., Joo, S., Kwigizile, V., & Oh, J. S. (2020). A pervasive framework toward sustainability and smart-growth: Assessing multifaceted transportation performance measures for smart cities. *Journal of Transport & Health, 19*, 100956. DOI: 10.1016/j.jth.2020.100956

Filho, G. P. R., Meneguette, R. I., Maia, G., Pessin, G., Gonçalves, V. P., Weigang, L., Ueyama, J., & Villas, L. A. (2020). A fog-enabled smart home solution for decision-making using smart objects. *Future Generation Computer Systems, 103*, 18–27. DOI: 10.1016/j.future.2019.09.045

Fu, X. (2024). Natural Language Processing in Urban Planning: A Research Agenda. *Journal of Planning Literature, 39*(3), 395–407. DOI: 10.1177/08854122241229571

Gkontzis, A. F., Kotsiantis, S., Feretzakis, G., & Verykios, V. S. (2024). Enhancing Urban Resilience: Smart City Data Analyses, Forecasts, and Digital Twin Techniques at the Neighborhood Level. *Future Internet 2024, Vol. 16, Page 47, 16*(2), 47. DOI: 10.3390/fi16020047

Golubchikov, O., & Thornbush, M. (2020). Artificial Intelligence and Robotics in Smart City Strategies and Planned Smart Development. *Smart Cities 2020, Vol. 3, Pages 1133-1144, 3*(4), 1133–1144. DOI: 10.3390/smartcities3040056

Grishina, A., Chinnici, M., Kor, A.-L., Rondeau, E., Georges, J.-P., & De Chiara, D. (2021). *Data Center for Smart Cities: Energy and Sustainability Issue*. 1–36. DOI: 10.1007/978-3-030-38836-2_1

Hao, H., Wang, Y., & Chen, J. (2024). Empowering Scenario Planning with Artificial Intelligence: A Perspective on Building Smart and Resilient Cities. *Engineering (Beijing), 43*, 272–283. Advance online publication. DOI: 10.1016/j.eng.2024.06.012

Hashem, I. A. T., Usmani, R. S. A., Almutairi, M. S., Ibrahim, A. O., Zakari, A., Alotaibi, F., Alhashmi, S. M., & Chiroma, H. (2023). Urban Computing for Sustainable Smart Cities: Recent Advances, Taxonomy, and Open Research Challenges. *Sustainability 2023, Vol. 15, Page 3916, 15*(5), 3916. DOI: 10.3390/su15053916

Hosseinian-Far, A., Ramachandran, M., & Slack, C. L. (2017). Emerging trends in cloud computing, big data, fog computing, IoT and smart living. In *Technology for Smart Futures* (pp. 29–40). Springer International Publishing., DOI: 10.1007/978-3-319-60137-3_2

Hussain, I. (2024). Secure, Sustainable Smart Cities and the Internet of Things: Perspectives, Challenges, and Future Directions. *Sustainability 2024, Vol. 16, Page 1390, 16*(4), 1390. DOI: 10.3390/su16041390

Javaid, S., Sufian, A., Pervaiz, S., & Tanveer, M. (2018). Smart traffic management system using Internet of Things. *International Conference on Advanced Communication Technology, ICACT, 2018-February*, 393–398. DOI: 10.23919/ICACT.2018.8323770

Kamrowska-Załuska, D., & Bibri, S. E. (2021). Impact of AI-Based Tools and Urban Big Data Analytics on the Design and Planning of Cities. *Land 2021, Vol. 10, Page 1209, 10*(11), 1209. DOI: 10.3390/land10111209

Keerthana, T., Kaviya, K., Deepthi Priya, S., & Kumar, A. S. (2021). AI enabled smart surveillance system. *Journal of Physics: Conference Series, 1916*(1), 12034. DOI: 10.1088/1742-6596/1916/1/012274

Khediri, A., Laouar, M. R., & Eom, S. B. (2021). Improving intelligent decision making in urban planning: Using machine learning algorithms. [IJBAN]. *International Journal of Business Analytics, 8*(3), 40–58.

Korada, L. (2021). Unlocking Urban Futures: The Role Of Big Data Analytics And AI In Urban Planning – A Systematic Literature Review And Bibliometric Insight. *Migration Letters : An International Journal of Migration Studies, 18*(6), 775–795. https://migrationletters.com/index.php/ml/article/view/11260

Köster, F., Ulmer, M. W., Mattfeld, D. C., & Hasle, G. (2018). Anticipating emission-sensitive traffic management strategies for dynamic delivery routing. *Transportation Research Part D, Transport and Environment, 62*, 345–361. DOI: 10.1016/j.trd.2018.03.002

Koumetio Tekouabou, S. C., Diop, E. B., Azmi, R., & Chenal, J. (2022). Artificial Intelligence Based Methods for Smart and Sustainable Urban Planning: A Systematic Survey. *Archives of Computational Methods in Engineering, 30*(2), 1421–1438. DOI: 10.1007/s11831-022-09844-2

Kourtit, K. (2021). City intelligence for enhancing urban performance value: A conceptual study on data decomposition in smart cities. *Asia-Pacific Journal of Regional Science, 5*(1), 191–222. DOI: 10.1007/s41685-021-00193-9

Li, J., Meng, Y., Ma, L., Du, S., Zhu, H., Pei, Q., & Shen, X. (2022). A Federated Learning Based Privacy-Preserving Smart Healthcare System. *IEEE Transactions on Industrial Informatics, 18*(3), 2021–2031. DOI: 10.1109/TII.2021.3098010

Li, S., Yuan, F., & Liu, J. (2024). Smart city VR landscape planning and user virtual entertainment experience based on artificial intelligence. *Entertainment Computing, 51*, 100743. DOI: 10.1016/j.entcom.2024.100743

Li, Z., Al Hassan, R., Shahidehpour, M., Bahramirad, S., & Khodaei, A. (2019). A Hierarchical Framework for Intelligent Traffic Management in Smart Cities. *IEEE Transactions on Smart Grid, 10*(1), 691–701. DOI: 10.1109/TSG.2017.2750542

Land Transport Authority (LTA). (2025). Intelligent Transport Systems. https://www.lta.gov.sg/content/ltagov/en/getting_around/driving_in_singapore/intelligent_transport_systems.html

Mahmoud, M., Perez-Ortiz, M., Asad Bokhari, S. A., & Myeong, S. (2022). Use of Artificial Intelligence in Smart Cities for Smart Decision-Making: A Social Innovation Perspective. *Sustainability 2022, Vol. 14, Page 620, 14*(2), 620. DOI: 10.3390/su14020620

Majeed, K., Jing, Y., Novakovic, D., & Ouazzane, P. K. (2014). Behaviour Based Anomaly Detection for Smartphones Using Machine Learning Algorithm. *International Conference on Computer Science and Information Systems.*

Mehmood, I., & Ning, H. (2024). Role of computer vision in smart cities: applications and research challenges. *Multimedia Tools and Applications 2024 83:5, 83*(5), 14883–14883. DOI: 10.1007/s11042-024-18276-y

Mortaheb, R., & Jankowski, P. (2023). Smart city re-imagined: City planning and GeoAI in the age of big data. *Journal of Urban Management, 12*(1), 4–15. DOI: 10.1016/j.jum.2022.08.001

Neelakandan, S., Berlin, M. A., Tripathi, S., Devi, V. B., Bhardwaj, I., & Arulkumar, N. (2021). IoT-based traffic prediction and traffic signal control system for smart city. *Soft Computing, 25*(18), 12241–12248. DOI: 10.1007/s00500-021-05896-x

New York City, D. O. T. (2025). Infrastructure: Traffic Signals. https://www.nyc.gov/html/dot/html/infrastructure/signals.shtml

Pharmacogn, J., Srilakshmi, A., Rakkini, J., Sekar, K. R., & Manikandan, R. (2018). A Comparative study on Internet Of Things (IoT) and its applications in Smart Agriculture. *Pharmacognosy Journal, 10*(2), 260–264. DOI: 10.5530/pj.2018.2.46

Qian, K., Mao, L., Liang, X., Ding, Y., Gao, J., Wei, X., Guo, Z., & Li, J. (2023). *AI Agent as Urban Planner: Steering Stakeholder Dynamics in Urban Planning via Consensus-based Multi-Agent Reinforcement Learning*. https://arxiv.org/abs/2310.16772v2

Rabby, M. K. M., Islam, M. M., & Imon, S. M. (2019). A review of IoT application in a smart traffic management system. *2019 5th International Conference on Advances in Electrical Engineering, ICAEE 2019*, 280–285. DOI: 10.1109/ICAEE48663.2019.8975582

Ranyal, E., Sadhu, A., & Jain, K. (2022). Road Condition Monitoring Using Smart Sensing and Artificial Intelligence: A Review. *Sensors 2022, Vol. 22, Page 3044, 22*(8), 3044. DOI: 10.3390/s22083044

Rathore, M. M., Paul, A., Ahmad, A., & Jeon, G. (2020). IoT-Based Big Data: From Smart City towards Next Generation Super City Planning. Https://Services.Igi-Global.Com/Resolvedoi/Resolve.Aspx?Doi=10.4018/978-1-5225-9866-4.Ch064, 1409–1428. DOI: 10.4018/978-1-5225-9866-4.ch064

Razavi, M., Hamidkhani, M., & Sadeghi, R. (2019). Smart Traffic Light Scheduling in Smart City Using Image and Video Processing. *Proceedings of 3rd International Conference on Internet of Things and Applications, IoT 2019*. DOI: 10.1109/IICITA.2019.8808836

Ren, Y., Leng, Y., Qi, J., Sharma, P. K., Wang, J., Almakhadmeh, Z., & Tolba, A. (2021). Multiple cloud storage mechanism based on blockchain in smart homes. *Future Generation Computer Systems, 115*, 304–313. DOI: 10.1016/j.future.2020.09.019

Rizwan, P., Suresh, K., & Rajasekhara Babu, M. (2017). Real-time smart traffic management system for smart cities by using Internet of Things and big data. *Proceedings of IEEE International Conference on Emerging Technological Trends in Computing, Communications and Electrical Engineering, ICETT 2016*. DOI: 10.1109/ICETT.2016.7873660

Sanchez, T. W., Fu, X., Yigitcanlar, T., & Ye, X. (2024). The Research Landscape of AI in Urban Planning: A Topic Analysis of the Literature with ChatGPT. *Urban Science 2024, Vol. 8, Page 197, 8*(4), 197. DOI: 10.3390/urbansci8040197

Satpathy, I., Nayak, A., & Jain, V. (2025). The Green City: Sustainable and Smart Urban Living Through Artificial Intelligence. Https://Services.Igi-Global.Com/Resolvedoi/Resolve.Aspx?Doi=10.4018/979-8-3693-6854-1.Ch009, 273–304. DOI: 10.4018/979-8-3693-6854-1.ch009

Seo, S. B., Yadav, P., & Singh, D. (2022). LoRa based architecture for smart town traffic management system. *Multimedia Tools and Applications*, *81*(19), 26593–26608. DOI: 10.1007/s11042-020-10091-5

Settimi, A., Gamerro, J., & Weinand, Y. (2022). Augmented-reality-assisted timber drilling with smart retrofitted tools. *Automation in Construction*, *139*, 104272. DOI: 10.1016/j.autcon.2022.104272

Sharma, A., Awasthi, Y., & Kumar, S. (2020). The Role of Blockchain, AI and IoT for Smart Road Traffic Management System. *Proceedings - 2020 IEEE India Council International Subsections Conference, INDISCON 2020*, 289–296. DOI: 10.1109/INDISCON50162.2020.00065

Shyamala Devi, M., Suguna, R., Joshi, A. S., & Bagate, R. A. (2019). Design of IoT Blockchain Based Smart Agriculture for Enlightening Safety and Security. *Communications in Computer and Information Science*, *985*, 7–19. DOI: 10.1007/978-981-13-8300-7_2

Skubis, I., Wolniak, R., & Grebski, W. W. (2024). AI and Human-Centric Approach in Smart Cities Management: Case Studies from Silesian and Lesser Poland Voivodships. *Sustainability 2024, Vol. 16, Page 8279, 16*(18), 8279. DOI: 10.3390/su16188279

Son, T. H., Weedon, Z., Yigitcanlar, T., Sanchez, T., Corchado, J. M., & Mehmood, R. (2023). Algorithmic urban planning for smart and sustainable development: Systematic review of the literature. *Sustainable Cities and Society*, *94*, 104562. DOI: 10.1016/j.scs.2023.104562

Tan, J., Yuan, Q., Guo, W., Xie, N., Liu, F., Wei, J., & Zhang, X. (2022). Deep Reinforcement Learning for Traffic Signal Control Model and Adaptation Study. *Sensors 2022, Vol. 22, Page 8732, 22*(22), 8732. DOI: 10.3390/s22228732

Tao, M., Qu, C., Wei, W., Zhou, B., & Huang, S. (2018). Hybrid Cloud Architecture for Cross-Platform Interoperability in Smart Homes. *Lecture Notes in Computer Science (Including Subseries Lecture Notes in Artificial Intelligence and Lecture Notes in Bioinformatics), 11336 LNCS*, 608–617. DOI: 10.1007/978-3-030-05057-3_45

Tran, T. V., Dang, N. T., & Chung, W. Y. (2017). Battery-free smart-sensor system for real-time indoor air quality monitoring. *Sensors and Actuators. B, Chemical*, *248*, 930–939. DOI: 10.1016/j.snb.2017.01.198

Tyagi, N., & Bhushan, B. (2023). Demystifying the Role of Natural Language Processing (NLP) in Smart City Applications: Background, Motivation, Recent Advances, and Future Research Directions. *Wireless Personal Communications*, *130*(2), 857–908. DOI: 10.1007/s11277-023-10312-8 PMID: 37168438

Vangala, A., Das, A. K., Kumar, N., & Alazab, M. (2020). Smart Secure Sensing for IoT-Based Agriculture: Blockchain Perspective. *IEEE Sensors Journal, 21*(16), 17591–17607. DOI: 10.1109/JSEN.2020.3012294

Verma, S., Gala, R., Madhavan, S., Burkule, S., Chauhan, S., & Prakash, C. (2018). An Internet of Things (IoT) Architecture for Smart Agriculture. *Proceedings - 2018 4th International Conference on Computing, Communication Control and Automation, ICCUBEA 2018*. DOI: 10.1109/ICCUBEA.2018.8697707

Wallace, R. B., Horsfall, F., Goubran, R., El-Haraki, A., & Knoefel, F. (2019, May 3). The Challenges of Connecting Smart Home Health Sensors to Cloud Analytics. *SAS 2019 - 2019 IEEE Sensors Applications Symposium,Conference Proceedings*. DOI: 10.1109/SAS.2019.8706128

Wang, H., Xu, P., & Zhao, J. (2021). Improved KNN Algorithm Based on Preprocessing of Center in Smart Cities. *Complexity, 2021*(1), 5524388. DOI: 10.1155/2021/5524388

Xiong, W., Wang, Y., Gao, S., Huang, X., & Wang, S. (2024). A multi-objective service composition optimization method considering multi-user benefit and adaptive resource partitioning in hybrid cloud manufacturing. *Journal of Industrial Information Integration, 38*, 100564. DOI: 10.1016/j.jii.2024.100564

Yao, H., Gao, P., Wang, J., Zhang, P., Jiang, C., & Han, Z. (2019). Capsule Network Assisted IoT Traffic Classification Mechanism for Smart Cities. *IEEE Internet of Things Journal, 6*(5), 7515–7525. DOI: 10.1109/JIOT.2019.2901348

Yassine, A., Singh, S., Hossain, M. S., & Muhammad, G. (2019). IoT big data analytics for smart homes with fog and cloud computing. *Future Generation Computer Systems, 91*, 563–573. DOI: 10.1016/j.future.2018.08.040

Yilmaz, O. N. C., Li, Z., Valkealahti, K., Uusitalo, M. A., Moisio, M., Lundén, P., & Wijting, C. (2014). Smart mobility management for D2D communications in 5G networks. *2014 IEEE Wireless Communications and Networking Conference Workshops. WCNCW, 219–223*, 219–223. Advance online publication. DOI: 10.1109/WCNCW.2014.6934889

Yu, J., Guo, L., Zhang, J., & Wang, G. (2024). A survey on graph neural network-based next POI recommendation for smart cities. *Journal of Reliable Intelligent Environments, 10*(3), 299–318. DOI: 10.1007/s40860-024-00233-z

Zheng, Y., Lin, Y., Zhao, L., Wu, T., Jin, D., & Li, Y. (2023). Spatial planning of urban communities via deep reinforcement learning. *Nature Computational Science 2023 3:9, 3*(9), 748–762. DOI: 10.1038/s43588-023-00503-5

Zhou, S., Wei, C., Song, C., Pan, X., Chang, W., & Yang, L. (2023). Short-Term Traffic Flow Prediction of the Smart City Using 5G Internet of Vehicles Based on Edge Computing. *IEEE Transactions on Intelligent Transportation Systems*, *24*(2), 2229–2238. DOI: 10.1109/TITS.2022.3147845

Chapter 5
Farmers' Rights in the Age of Agricultural Intellectual Property:
From Soil to Statute

Ananya Pandey
https://orcid.org/0000-0002-2419-6314
Christ University, Bangalore, India

Achyutananda Mishra
https://orcid.org/0000-0002-4113-0974
Christ University, Bangalore, India

ABSTRACT

This chapter explores the evolving landscape of Farmers' Rights in the context of intensifying agricultural intellectual property regimes. It critically examines the intersection of international legal frameworks, national legislations, and customary practices that shape farmers' access to seeds, biodiversity, and knowledge systems. Through a multidisciplinary lens combining legal theory, environmental justice, and agroecological insights, the chapter interrogates how intellectual property regimes, such as UPOV and TRIPS, impact traditional seed systems, biodiversity conservation, and food sovereignty. It highlights tensions between privatized biotechnological innovations and community-centric stewardship of genetic resources, particularly in the face of climate change and digital sequence information. By tracing the trajectory from soil to statute, farmers' rights emerge from grassroots agricultural practices and struggles, evolving into legal recognition and protection through policy and law.

DOI: 10.4018/979-8-3373-3206-2.ch005

INTRODUCTION

The accelerating transformation of global agricultural systems under the influence of intellectual property regimes, biotechnological innovations, and market-driven legal architectures has precipitated a profound conflict between traditional agrarian knowledge and modern frameworks of seed ownership (Brush, 2005; Kloppenburg, 2010). In this evolving context, farmers' rights have emerged as a central focal point of attraction in the discourse on sustainable agriculture, biodiversity conservation, and equitable development (Andersen, 2008; De Schutter, 2009). These rights—rooted in the recognition of farmers as custodians of plant genetic resources and bearers of intergenerational ecological knowledge—are fundamental to ensuring food sovereignty, environmental sustainability, and social justice (FAO, 2009; Girsberger, 1999). While the International Treaty on Plant Genetic Resources for Food and Agriculture (ITPGRFA) enshrines these rights in international law (FAO, 2009), their implementation remains deeply contested, particularly where they intersect with the more dominant and commercially incentivized structures of agricultural intellectual property law, such as the Agreement on Trade-Related Aspects of Intellectual Property Rights (TRIPS) and the International Union for the Protection of New Varieties of Plants (UPOV)Conventions (Correa, 2000; Dutfield, 2011). In many jurisdictions, intellectual property protection mechanisms have expanded to encompass plant varieties, genetically modified organisms (GMOs), and increasingly, digital sequence information (DSI), often privileging corporate innovation over the traditional, collective, and informal seed systems that smallholder farmers have relied upon for centuries (Halewood et al., 2013; Jefferson et al., 2020; Louafi, 2022). The "soil to statute" perspective adopted here underscores the dynamic interface between lived agricultural practices and the codified legal architectures that govern them (Carpa & Mattei, 2015). By doing so, it aims to unravel how farmers' rights are simultaneously constituted, constrained, and contested within a rapidly changing the legal landscape. In light of the above perspective, this chapter explores the legal landscape of farmers' rights, a framework that situates law within the broader interdependencies of ecological systems, sociocultural contexts, and political economies

AGRICULTURAL INTELLECTUAL PROPERTY

An important part of a nation's economy is agriculture, both in terms of creating jobs and its GDP contribution. Almost half of the developing state's population (like India and Brazil) works in agriculture, which also contributes significantly to its economic production (World Bank group, 2025). Food grains and cereals, fruits, vegetables, and a number of commercial crops like oilseeds, cotton, rubber, spices, sugar cane, jute, and tobacco are all included in this broad sector (Horton, 1998). The innovation and development in agricultural sector are highly dependent with the protection granted to plant varieties under the intellectual property regime (Peschard, 2017). IPR in agriculture mostly deals with patents, plant breeder's rights, trademarks, geographical indications, and trade secrets and is used to protect products or services produced in the agricultural sector (Campi & Nuvolari, 2021). Agricultural innovations include creating food for people and animals that is safer, healthier, and more nutrient-dense (e.g., genetically modified seeds, novel breeding techniques) (Campi, 2017). Productivity in the agriculture industry is driven by innovation. Nevertheless, there are ambiguities in the Act that allow others to steal varieties and create inferior seeds, which causes farmers' crops to fail (Delmer, 2003).

Recent decades have seen some incredible advancements in agriculture, particularly the introduction of advantageous features into numerous crops that would otherwise be extinct as a result of pests, illnesses, and droughts (Borowiak, 2004). Crops can produce higher yields with the aid of resistant seed variants while using fewer chemical fertilizers. It is especially beneficial to farmers in emerging nations like India (Oguamanam, 2014). The application of this aspect of agricultural innovation can significantly aid in the reduction of poverty, hunger, food security, and illness. Agricultural innovations are guided, supported, created, transferred, or adopted by a wide range of stakeholders, including farmers, researchers, private companies, advisers, non-governmental organizations (NGOs), consumers, and others (Campi, 2017). Farmers and the general public are also advised and informed about these innovations. For instance, in India, the significance of safeguarding intellectual property in relation to agricultural innovation has increased over the past several decades (Campi & Nuvolari, 2021). Although investing in innovation is a key factor in economic growth, governments face limitations in funding innovative projects, such as agricultural research and development (Prasad et al., 2023). Crop innovation encompasses five phases: discovery, proof of concept, early development, advanced development, and pre-launch. Commercially viable seeds can take 10 to 15 years to develop (Jewell, 2015). Thus, investors can engage in long-term investments in innovation due to patent protection and adherence to regulations. As a result, IPRs have garnered more focus for their role in promoting agricultural development, encompassing aspects such as foreign direct investment (FDI), technology transfer, trade, access to genetic resources, and safeguarding traditional knowledge (Curtis, 2012).

Concept of Soil to Statute

The concept of "soil to statute" covers the wide intersection of the land resources particularly plants and seeds with the legal landscape governing their ownership, usage and conservation. Soil i.e., the foundational and fundamental for seeds and plants hold immense socio-economic and legal value (FAO, 2015). It represents a symbolic heritage and identify that is held by farmers and is of profound significance. Soil is the root from which the innovation and development in the form of plant and seed varieties take place (FAO, 2010). The development in the form of agricultural sector with the support of intellectual property, as discussed above, has tried to increasingly influence and undermine the seed system that raises a significant conflict with the traditional practice rooted in the soil (Clancy & Moschini, 2017). Statutory protection in developing countries like India in the form of Plant Variety Protection and Farmers Rights (PPV&FRA), 2001 connects the protection of plant in the form of plant varieties in a concrete legal form (Peschard, 2017). Furthermore, in the developing countries particularly, soil is considered to be a metaphor for sovereignty, where the seed control and farming practice are all subject to, under control of the local food system. Often the statutory framework also fails to recognize the farmers contribution and that is threatening the centuries of ecological and traditional knowledge that is cultivated by farmers with their knowledge, experience and hard work (Stringer et al., 2020). For this to fall in place, the statute has to go beyond recognition where the farmers rights are subject to active and exclusive protection (FAO, 2015). Therefore, the interplay between soil to statutory framework under the realm of plant protection emphasizes the need for safeguard farmers interests that is connected and embedded in the soil (FAO, 2010).

FARMERS RIGHTS: CONCEPTUAL AND LEGAL FOUNDATION

In order to draw attention to the important but underappreciated contributions farmers make to plant genetic resources for food and agriculture, Pat Roy Mooney (Mooney, 2011) and Cary Fowler of the then-Rural Advancement Foundation International (now ETC-Group) first used the term "farmers' rights" as a political concept in the early 1980s (Farmers Rights, 2022). Significant clarifications have been made since 1986, when Farmers' Rights were first brought up in international discussions. One of the main issues in the negotiations has been how to compensate farmers for their past, present, and future contributions to the preservation, enhancement, and accessibility of crop genetic resources for agriculture and food. It is helpful to read over the history of discussions on farmers' rights, the conceptual and legal foundations in order to comprehend the provisions on these rights in the international treaty.

Conceptual Foundations

At the core of contemporary agricultural governance lies a complex constellation of legal norms and frameworks that intersect across the domains of intellectual property rights, environmental law, and human rights. Among these, the concept of Farmers' Rights has emerged as both a legal and moral imperative in the struggle to protect biodiversity, secure livelihoods, and ensure food sovereignty. Originating from the recognition of the indispensable role of farmers—particularly smallholders, Indigenous peoples, and local communities—in conserving and developing plant genetic resources for food and agriculture (PGRFA), Farmers' Rights are formally articulated in key international legal instruments discussed below. The ITPGRFA, adopted by the Food and Agriculture Organization (FAO) in 2001, provides the most comprehensive global recognition of Farmers' Rights. Article 9 of the ITPGRFA mandates states to protect and promote Farmers' Rights, which include the protection of traditional knowledge relevant to PGRFA, the right to equitably participate in benefit-sharing arising from the utilization of these resources, and the right to participate in decision-making processes (FAO, 2009). Importantly, the ITPGRFA leaves the responsibility of implementation to national governments, which has led to uneven and often inadequate enforcement (Andersen, 2008).

Complementary to the ITPGRFA is the CBD, adopted in 1992, which enshrines the principles of access and benefit-sharing (ABS), traditional knowledge protection, and the sovereign rights of states over their natural resources. While the CBD does not explicitly refer to Farmers' Rights, its provisions regarding community participation and equitable benefit-sharing provide a foundational legal basis for supporting them (CBD, 1992). Additionally, the 2018 United Nations Declaration on the Rights of Peasants and Other People Working in Rural Areas (UNDROP) offers a rights-based articulation of farmers' entitlements, affirming their rights to seeds, land, biodiversity, and participation in decision-making (United Nations, 2018). Taken together, these instruments underscore a shift in the international legal order—away from purely market-oriented models of agriculture and toward inclusive frameworks that acknowledge the collective, intergenerational, and ecological dimensions of farming (Shiva, 2000& Baldwin, 2009).

The dominant legal narratives underpinning agricultural development continue to be shaped by intellectual property rights (IPRs), which create enforceable claims over plant varieties and genetic materials, often at the expense of traditional farming practices (Gibson, 2004). The expansion of the Agreement on TRIPS, enforced by the World Trade Organization (WTO), has accelerated the global diffusion of IPR standards, requiring member states to provide protection for plant varieties either through patents or an effective sui generis system (Correa, 2000). While TRIPS permits flexibility in national implementation, in practice this flexibility is often constrained by bilateral trade agreements and the influence of transnational seed corporations (Dutfield, 2008).

The UPOV—particularly its 1991 Act—exemplifies a more restrictive form of plant variety protection (PVP) that grants breeders exclusive rights over new, distinct, uniform, and stable (DUS) varieties. UPOV-91 severely limits farmers' ability to save, reuse, exchange, or sell farm-saved seeds—practices that have historically underpinned agricultural biodiversity and resilience (De Jonge, 2014). Although UPOV is often promoted as incentivizing innovation and investment in plant breeding, its incompatibility with Farmers' Rights has been widely critiqued. By legally privileging formal breeding institutions and multinational seed companies, UPOV marginalizes the informal seed systems on which the vast majority of the world's smallholder farmers rely (La Vía Campesina, 2015). In contrast, countries such as India have developed sui generis regimes like the Protection of Plant Varieties and Farmers' Rights Act (PPV&FR Act) of 2001, which attempt to balance breeders' rights with legally enforceable Farmers' Rights, including the right to save, use, sow, resow, exchange, share, or sell farm produce including seed (Sahai, 2003).

Moreover, the proliferation of biotechnological patents—covering genetically modified organisms (GMOs), gene sequences, and synthetic biology components—has introduced new complexities into the intellectual property landscape (Mooney, 2011). Patents confer stronger and broader monopolies than PVPs, often covering traits that are essential for food security, such as drought resistance or nutrient fortification (RAFI, 1998). The proprietary nature of these innovations, coupled with the legal enforcement of material transfer agreements and technology use restrictions, has generated significant concern about farmers' dependency, loss of sovereignty, and the erosion of local innovation systems (Kloppenburg, 2010). This legal enclosure of genetic commons not only exacerbates inequality between the Global North and South but also poses a threat to agroecological sustainability by promoting uniform, input-intensive, and proprietary cropping systems (Chaudhary et al., 2019).

Consequently, the agricultural intellectual property landscape presents a paradox. On one hand, it promises technological advancement, yield improvement, and climate resilience. On the other hand, it undermines the very systems—traditional knowledge, biodiversity, and informal seed networks—that are essential for sustainable agriculture. Navigating this paradox requires a reconceptualization of innovation, ownership, and rights within agriculture. It calls for legal models that transcend the binary of private versus public ownership and recognize the pluralistic, overlapping, and often communal nature of seed knowledge and use. Legal landscape thus provides analytical tools to interrogate not only the normative content of laws but also the political economies that shape their creation, interpretation, and implementation.

International Legal Framework

The historical trajectory of farmers' rights is intricately interwoven with the evolution of agricultural knowledge systems, colonial interventions, and the development of intellectual property regimes. Traditional agrarian communities across the Global South have long practiced seed saving, seed sharing, and crop innovation through community-based knowledge systems that were inherently collective, adaptive, and ecologically embedded (Shiva, 1997; Brush, 2005). These practices constituted the cornerstone of sustainable farming, where the seed was not a commodity but a shared resource and symbol of life. Over generations, indigenous and local farmers refined and passed down plant genetic knowledge, selectively breeding crops to adapt to specific climatic, topographical, and ecological conditions (Cleveland & Soleri, 2007). These farmers were not merely agriculturalists but innovators and custodians of agrobiodiversity. Their role in conserving and enriching genetic resources created the foundation of modern plant breeding (Andersen, 2008). However, this traditional system of knowledge and innovation was neither formally recognized by legal systems nor protected under conventional intellectual property frameworks, which have historically privileged individual invention over collective innovation (Correa, 2000).

Colonial and post-colonial intellectual property regimes significantly disrupted and devalued these indigenous systems of agricultural knowledge. During the colonial period, European empires-imposed state-centric legal systems that facilitated the appropriation of biological resources and displaced customary land and resource rights (Raj, 2007; Kloppenburg, 2004). Colonized territories were often treated as vast repositories of germplasm, and seeds were collected, catalogued, and shipped to botanical gardens and research institutions in the Global North (Kingsbury, 2009). This process of extraction laid the groundwork for later bioprospecting and biopiracy (Posey & Dutfield, 1996). Colonial agricultural policies typically imposed monoculture systems aimed at export economies, thereby undermining local food sovereignty and biodiversity (Shiva, 2016). Post-colonial legal infrastructures inherited these colonial frameworks and continued to marginalize the informal seed systems maintained by local communities (Chaudhary, 2013). While many post-independence states attempted to modernize agriculture through national seed laws, research institutes, and subsidies, they often replicated hierarchical, technocratic models that failed to account for farmers' experiential knowledge and ecological wisdom (Louwaars, 2007). As a result, a dual legal system emerged: one formal and codified, governing scientific innovation and commercial agriculture, and another informal and often invisible, sustaining the real-life practices of millions of smallholder farmers. The marginalization of customary norms within state and international legal frameworks remains a core challenge in asserting and operationalizing farmers' rights today (De Schutter, 2009).

The transition from collective rights to commercial patents represents a paradigmatic shift in the governance of agricultural knowledge and resources. Beginning in the mid-20th century, particularly with the rise of industrial agriculture and biotechnology, the seed became increasingly commodified (Howard, 2009). The introduction of plant variety protection (PVP) laws, such as those governed by the International Union for the Protection of New Varieties of Plants (UPOV), introduced proprietary rights over plant varieties, allowing breeders to restrict the use, reuse, and exchange of seeds (Leskien & Flitner, 1997). These laws were designed around the model of formal plant breeding institutions and private sector innovation, largely sidelining the informal innovation systems of farmer communities (Louwaars & De Boef, 2012). The Trade-Related Aspects of Intellectual Property Rights (TRIPS) Agreement under the World Trade Organization (WTO) further internationalized this approach, mandating member countries to adopt minimum standards of intellectual property protection for plant varieties—either through patents or effective sui generis systems (WTO, 1994). While TRIPS allowed some flexibility, its implementation has overwhelmingly favored UPOV-style protections that undermine traditional seed systems (GRAIN, 2005).

This legal shift has profound implications for farmers' rights, particularly their right to save, use, exchange, and sell farm-saved seed. Once a fundamental aspect of agricultural practice, these rights are increasingly restricted or criminalized under intellectual property laws (Andersen, 2008). The commercial seed market, dominated by a handful of multinational corporations, often requires farmers to sign contracts that prohibit seed reuse, enforce royalty payments, and limit autonomy (ETC Group, 2013). These contracts, while legally enforceable, are often not fully understood by resource-poor farmers and exacerbate existing power asymmetries (Kloppenburg, 2010). The privatization of seed systems also correlates with a loss of agrobiodiversity, as corporations prioritize high-yield, genetically uniform varieties over diverse, locally adapted landraces (FAO, 2010). The result is a narrowing of the genetic base, reduced ecological resilience, and increased vulnerability to climate change and pests.

Despite these developments, resistance to the commercial patent regime is growing, particularly through the assertion of farmers' rights at national and international levels. Legal frameworks such as India's Protection of Plant Varieties and Farmers' Rights Act (PPV&FR Act) represent an attempt to reconcile the competing imperatives of intellectual property protection and farmers' entitlements (Rout, 2012). This law uniquely recognizes the role of farmers as breeders and conservers of biodiversity, granting them specific rights that go beyond the scope of UPOV. However, even such progressive laws face implementation challenges, including bureaucratic inertia, lack of awareness, and conflicts with existing seed policies (Sahai, 2003). Globally, instruments like the International Treaty on Plant Genetic Resources for Food and Agriculture (ITPGRFA) have attempted to institutionalize farmers' rights through provisions for benefit-sharing, participation in decision-making, and protection of traditional knowledge (FAO, 2009). Yet, these instruments often lack enforceability and remain aspirational rather than operational (Visser et al., 2005).

The historical evolution of farmers' rights thus reveals a tension between tradition and modernity, between communal innovation and private ownership, and between ecological stewardship and market-driven commodification. It underscores the need for a legal landscape that recognizes the pluralism of knowledge systems and the legitimacy of farmer-centered innovation. Addressing this requires not only legal reform but also a paradigm shift in how we conceptualize innovation, ownership, and the commons in agriculture. As we move into an era of digital agriculture, synthetic biology, and global food insecurity, learning from this historical evolution becomes imperative for designing equitable and sustainable intellectual property regimes that genuinely empower farmers and protect agrobiodiversity.

National Legal Framework

The fundamental legal basis for the protection of farmers rights is the WTO's TRIPS Agreement. All WTO members are required by Article 27.3(b) to safeguard plant varieties by patents, an efficient sui generis system, or any combination of these (Verma, 2010). The Agreement does not, however, specify what is meant by "sui generis" or "effective," therefore it is unclear if and to what degree it permits a national IPR system to grant farmers the right to preserve, use, trade, and/or sell farm-saved seed (Prip. 2024). The seed industry believes that Plant Breeders' Rights (PBR) under the UPOV establish the only appropriate standards for implementing TRIPS Article 27.3(b), whereas the sui generis flexibility was employed to placate the numerous protesters against the mandatory IPR protection of plant types (Borowiak, 2004). Various regions protect plant varieties in different forms of statutes, policies and legal framework. Some of them have been discussed below:

Brazil

Before Brazil joined the WTO, its laws did not offer intellectual property protection for plant varieties. Although a Seed Bill passed in 1965 set standards for seed production and trade, it did not include specific protections for breeders' rights (Migliori, 2021). The public sector was significantly involved in plant breeding, and the new cultivars it produced were kept in the public domain (Peschard, 2017). Private seed companies proliferated, engaged in seed distribution, and focused on breeding crops suitable for hybridization, like corn. Brazil reformed its plant variety legislation by amending the Industrial Property Act and enacting the Plant Variety Protection (PVP) Act after the TRIPS Agreement came into effect (Semon, 2003). The PVP Act came into force in 1997. The Act created an important legal reference point by introducing IP rights for plant varieties in Brazil. According to the PVP Act, "The protection of intellectual property rights in plant varieties is achieved through the issuance of a Plant Variety Protection Certificate (Peschard, 2017). This certificate is regarded as a commodity for all legal purposes and serves as the only form of protection for plant varieties and their associated rights against the free use of sexually or vegetatively propagated plants or parts thereof (Semon, 2003)."

European Union

The EU has set up a mechanism known as Community plant variety right (CPVR) that confers intellectual property rights on new plant varieties (Ghijsen, 2002). It resembles a patent and, once granted, is applicable across the EU. It complies with the TRIPS/WTO agreements and the UPOV 1991 convention (Kiewiet, 2005). Located in Angers, France, the Community Plant Variety Office (CPVO) carries out the system. When breeders apply to the CPVO, they obtain a single intellectual property right that is valid across the entire EU (Velasco, 2020). Most plant species can be granted plant breeders' rights for 25 years. Vine species, trees, flower bulb groups, small woody fruits, ornamental woods, potatoes, and asparagus can be protected for 30 years (Ghijsen, 2002). The CPVO does not depend on EU institutions for its finances. It makes decisions regarding plant variety protection titles that are legally binding for third parties (Llewelyn & Adcock, 2006).

India

Before India became a member of the WTO, it did not offer intellectual property protection for plant varieties (Venkatesan, 2018). The initial version of its plant variety protection bill, put forward in 1993, faced significant resistance and led to widespread farmer protests known as beej satyagraha, or "seed protest" (Peschard & Randeria, 2020). Consequently, revised drafts were presented in 1997, 1999, and 2000. A Joint Parliamentary Committee conducted public hearings across India from January to August 2000 (Tonapi, 2008). The Protection of Plant Varieties and Farmers' Rights (PPVFR) Act was finally enacted in 2001 after a struggle lasting seven years and involving five drafts. India is notable as one of the few countries globally that has enacted sui generis legislation for plant variety protection (Winge et al., 2013).

United States of America

The Plant Variety Protection Act (PVPA) of 1970 was designed to motivate seed companies to enhance crop varieties beyond hybrids. Breeders could secure a Plant Variety Protection certificate (PVPC) under the act to safeguard their intellectual property rights for new varieties (Fuglie & MacDonald, 2023). Farmers were permitted to save seeds from varieties that had certificate protection, but selling bin-run seed to other farmers required a license from the breeding company that owned the certificate (Alston & Venner, 2002). Nevertheless, other seed companies and breeders were allowed to use protected varieties as parent material in their own breeding programs without restriction. While those protections did lead to some private research and development, it varied across different crops (Chen, 2005).

CONFLICT BETWEEN TRADITIONAL AGRARIAN KNOWLEDGE AND MODERN FRAMEWORK OF SEED OWNERSHIP

The evolving interface between intellectual property (IP) regimes and Farmers' Rights reflects a deep-rooted tension between commodification and the commons in agricultural governance. At the heart of this contention lies the divergence between the International Union for the Protection of New Varieties of Plants (UPOV), particularly its 1991 Act, and the International Treaty on Plant Genetic Resources for Food and Agriculture (ITPGRFA). While UPOV advances a breeder-centric paradigm grounded in exclusive proprietary rights over plant varieties (UPOV, 1991), the ITPGRFA enshrines a framework of farmers' entitlements to conserve, use, exchange, and sell seeds, especially those developed through traditional knowledge systems (FAO, 2009). The UPOV 1991 framework severely limits farmers' autonomy by restricting the re-use of farm-saved seeds, often compelling them to rely on commercial seed markets and licensing agreements (GRAIN, 2015). In contrast, Article 9 of the ITPGRFA articulates Farmers' Rights as essential for the realization of agrobiodiversity and global food security, recognizing their historical role as stewards and innovators (FAO, 2009). This normative dissonance poses significant implications for seed sovereignty, particularly in countries of the Global South, where agricultural biodiversity is intricately tied to cultural heritage and local food systems (Andersen, 2008; Correa, 2000). The asymmetry in enforcement mechanisms and institutional authority between UPOV and ITPGRFA further exacerbates the imbalance, as IP-centric instruments enjoy more robust international and trade-based enforcement through mechanisms such as the WTO's TRIPS Agreement (WTO, 1994), whereas Farmers' Rights remain largely aspirational and soft-law commitments, often subject to national discretion (De Jonge, 2011).

National legislations serve as the arena where these competing frameworks collide and coalesce, offering a spectrum of legal responses to reconcile or reinforce such tensions. India's Protection of Plant Varieties and Farmers' Rights (PPV&FR) Act, 2001, is often lauded as a pioneering statute that attempts to operationalize a sui generis system accommodating both breeders' and farmers' rights (Sahai, 2003). However, a critical examination reveals several structural and functional shortcomings. On the surface, the Act appears to offer robust protection to farmers by granting them the right to save, use, sow, re-sow, exchange, and even sell farm produce, including seed, except when it involves branded seeds of protected varieties (PPV&FR Act, 2001). Yet, the distinction between branded and non-branded seeds remains ambiguous and susceptible to legal interpretation, potentially curtailing farmers' rights in practice (Ramanna, 2006). Furthermore, the procedural complexity and institutional inaccessibility embedded in the application and registration processes under the Act often place farmers—especially smallholders and indigenous cultivators—at a disadvantage compared to corporate breeders who are better resourced and legally equipped (Rout, 2011). The implementation of benefit-sharing provisions under the Act has also been sporadic, marred by bureaucratic delays and a lack of clarity on criteria and distribution mechanisms (Chiarolla & Louafi, 2015). Unlike UPOV-compliant models adopted in several countries, which prioritize breeders' commercial interests and afford limited flexibility for traditional seed exchange practices, India's sui generis approach is unique but remains compromised by internal contradictions and lack of operational coherence (Prakash, 2017). Countries like Kenya and the Philippines, which have moved toward UPOV-aligned legislation under external pressures such as trade agreements, demonstrate how compliance with IP-centric norms can undermine pre-existing community-based seed systems and marginalize local farming communities (Louwaars et al., 2005; GRAIN, 2015). The comparative analysis underscores that while national discretion offers theoretical space for integrating Farmers' Rights, the policy and institutional context often determine the extent and efficacy of such integration (Brush, 2005).

Overlaying this statutory terrain is the complex dynamic between customary law and formal legal systems, particularly in agrarian communities where seed knowledge and exchange operate within culturally embedded norms. Customary practices such as communal seed banks, intergenerational knowledge transmission, and informal seed networks remain vital for sustaining agrobiodiversity, yet they often fall outside the purview of statutory protection (Posey, 1996). The imposition of formal intellectual property laws on such communities can create legal dissonance, criminalizing traditional practices and disrupting socio-ecological systems (Andersen & Winge,

2012). Legal pluralism, therefore, emerges not just as a descriptive reality but as a normative imperative for ensuring justice in the governance of plant genetic resources. Recognizing and integrating community-based rights into statutory frameworks requires more than token acknowledgment; it demands institutional mechanisms that can harmonize plural legal orders without subordinating one to the other (Boer, 2016). Models of community protocols, participatory legal drafting, and customary tribunals offer promising avenues but remain underutilized in mainstream intellectual property governance (Wynberg & Chennells, 2009). In reconciling the tension between formal law and customary norms, a pluralistic legal landscape that centers the lived realities and epistemologies of farming communities is essential. Such an approach would not only restore balance between innovation and tradition but also pave the way for a more equitable and sustainable global food system.

ENVIRONMENT AND BIODIVERSITY IMPLICATIONS

The convergence of intellectual property (IP) rights and modern agriculture has precipitated a fundamental shift in seed governance, with profound consequences for environmental integrity and agrobiodiversity. The expansion of intellectual property regimes—particularly patents and plant breeders' rights—has facilitated the proliferation of genetically uniform, proprietary seed varieties, often at the expense of traditional, diverse landraces cultivated by farmers over generations (Kloppenburg, 2010; Moeller & Stannard, 2013). This homogenization of seeds, a hallmark of industrial agriculture, has engendered monocultures that pose significant ecological risks. Monocultural systems are inherently vulnerable to pests, diseases, and climatic variability due to their genetic uniformity, which undermines the resilience of agricultural ecosystems (Altieri, 2009). These vulnerabilities have, in turn, fostered increased reliance on agrochemicals and mechanized interventions, contributing to soil degradation, water pollution, and greenhouse gas emissions (Shiva, 2016; FAO, 2019). From an environmental law perspective, IP-driven monocultures are not merely a matter of technological preference but represent a legal architecture that systematically sidelines ecological considerations in favor of market-driven imperatives (Cullet, 2005).

In contrast, agrobiodiversity—defined as the variety and variability of plant species used in agriculture—is intrinsically linked to the practices of smallholder and indigenous farmers who conserve, improve, and share seeds as a cultural and ecological commons (Brush, 2005; Anderson, 2013). These farmer-managed systems are pivotal in sustaining ecological resilience by maintaining genetic diversity within and across crop species, which serves as a buffer against environmental shocks and long-term climatic change (FAO, 2021). The custodianship of agrobiodiversity by local communities exemplifies a dynamic interplay between cultural heritage and ecological stewardship, which stands in stark contrast to the proprietary logic of intellectual property regimes (Chiarolla, 2011). However, the prevailing legal frameworks often fail to recognize or protect this collective knowledge system. While instruments like the International Treaty on Plant Genetic Resources for Food and Agriculture (ITPGRFA) affirm the role of farmers in conserving genetic resources (FAO, 2009), the enforcement of such rights remains fragmented and subordinated to IP-based commercial interests (De Schutter, 2009).

Climate change further complicates the environmental implications of agricultural intellectual property by intensifying the demand for resilient seed varieties capable of withstanding extreme weather, shifting seasons, and emerging pests (Jarvis et al., 2016). In this context, the role of farmers as innovators and adaptors becomes critical. Yet, restrictive intellectual property regimes can inhibit farmers' access to seeds, limit their rights to save and exchange them, and disincentivize local breeding efforts (Louwaars et al., 2009). Legal frameworks that prioritize proprietary innovation over ecological adaptation risk stifling the grassroots innovations necessary for climate-resilient agriculture (Peschard & Randeria, 2020). Moreover, the regulatory gap surrounding emerging biotechnologies—such as CRISPR and synthetic biology—raises concerns about future access to genetic resources and the erosion of farmers' agency (Jasanoff et al., 2021). Therefore, integrating environmental sustainability into intellectual property governance is not merely an ecological necessity but a legal imperative. A legal landscape approach advocates for pluralistic frameworks that balance IPRs with environmental law, farmers' rights, and biodiversity conservation (Cullet & Mayet, 2011). Reimagining seed governance through this lens entails not only legal reforms but also a paradigm shifts toward agro ecological resilience, equity, and climate justice. Protecting agrobiodiversity and enabling farmer-driven innovation must become foundational pillars in the design of intellectual property regimes that serve ecological and public interest goals.

SOCIO-ECONOMIC IMPACT ON FARMERS

The socioeconomic dimensions of farmers' rights are deeply embedded in the legal structures governing Access and Benefit Sharing (ABS) mechanisms, as articulated under the Convention on Biological Diversity (CBD) and the International Treaty on Plant Genetic Resources for Food and Agriculture (ITPGRFA) (Convention on Biological Diversity, 1992; FAO, 2009). These instruments aim to rectify historical imbalances by mandating fair and equitable sharing of benefits derived from the utilization of genetic resources, with a particular emphasis on the role of traditional knowledge and community innovation (De Jonge, 2011). However, in practice, the implementation of ABS frameworks has often struggled to translate legal principles into meaningful economic empowerment for smallholder farmers and indigenous communities (Nijar, 2011). The transactional nature of benefit-sharing agreements tends to prioritize state-level negotiations, frequently sidelining the local actors who are the primary custodians of agrobiodiversity (Chiarolla, 2012). This disconnect is further exacerbated by power asymmetries in the global agricultural sector. The consolidation of agribusiness corporations, exemplified by mergers among biotech giants, has led to a monopolization of seed markets and the patenting of plant varieties—effectively marginalizing farmers' access to germplasm and limiting their ability to save, exchange, or develop seeds (Howard, 2015; Kloppenburg, 2010). Intellectual property regimes, particularly those aligned with the UPOV Convention and TRIPS Agreement, often reinforce these imbalances by privileging commercial breeders over traditional seed keepers (Correa, 2013). The result is a structural entrenchment of inequality, where legal and economic leverage is concentrated in the hands of a few global entities, while the majority of resource-dependent farmers remain disempowered (Coombe, 2009). Moreover, the intersectional dimensions of gender, indigeneity, and local epistemologies are frequently overlooked in policy frameworks, further entrenching socioeconomic disparities (Pimbert, 2009). To uphold the spirit of farmers' rights, there is an urgent need for ABS regimes and intellectual property laws to adopt a more inclusive, equity-centered approach that recognizes and supports the vital contributions of farming communities to global food security and biodiversity conservation.

EMERGING TRENDS AND CHALLENGES

The contemporary legal landscape surrounding Farmers' Rights is increasingly shaped by emerging technological developments and accompanying governance challenges. One significant frontier is Digital Sequence Information (DSI), which refers to the digital representation of genetic resources, often extracted from biological material without direct access to physical specimens (Laird & Wynberg, 2018). While DSI holds immense potential for scientific innovation and open-source collaboration, it also raises complex questions about data sovereignty, equitable benefit-sharing, and intellectual property (IP) protection (Bagley, 2020). Many low- and middle-income countries express concern that DSI enables bio-digital prospecting—where commercial entities use open databases to access genetic data without formal agreements or obligations under the Convention on Biological Diversity (CBD) or the Nagoya Protocol (Scholz et al., 2020). Consequently, DSI threatens to decouple genetic resources from their traditional custodians, eroding the core of Farmers' Rights in the digital age (Pauchard & Godt, 2021).

Concurrently, breakthroughs in CRISPR and synthetic biology present both opportunities and regulatory dilemmas. These genome-editing tools allow for precise modifications without introducing foreign DNA, blurring the definitional boundaries of what constitutes a genetically modified organism (GMO) (Montoliu, 2019). The absence of a cohesive global regulatory framework for such technologies creates a legal vacuum, enabling biotech companies to patent gene-edited seeds that may build upon indigenous germplasm or farmer-bred varieties, often without consent or compensation (Jasanoff et al., 2015). This undermines farmers' control over their genetic heritage and amplifies existing asymmetries in access, innovation, and legal recourse (Chhetri et al., 2022).

Lastly, the fragmentation of global governance frameworks exacerbates these challenges. Divergent interpretations of Farmers' Rights, intellectual property obligations, and benefit-sharing protocols across treaties like TRIPS, ITPGRFA, CBD, and UPOV create jurisdictional overlaps and normative conflicts (Louafi, 2020). Without harmonized legal instruments or inclusive enforcement mechanisms, farmers remain at a disadvantage within an increasingly complex and corporatized agro-biotechnological regime. Addressing these gaps requires not only legal coherence but a re-centering of farmers as agents of innovation, sustainability, and justice (Andersen, 2016).

TOWARDS A POLICY APPROACH

In envisioning a reimagined legal landscape for Farmers' Rights, it is imperative to adopt a pluralistic legal model that harmonizes statutory, customary, and international legal norms. The fragmentation of legal instruments governing agricultural biodiversity—ranging from the Convention on Biological Diversity (CBD) and the International Treaty on Plant Genetic Resources for Food and Agriculture (ITPGRFA) to national seed laws and intellectual property regimes—has created normative tensions that disproportionately affect smallholder and indigenous farming communities (Correa, 2013; Andersen, 2008). A pluralistic approach acknowledges the legitimacy of local customary systems that govern seed exchange and biodiversity conservation, allowing these to coexist with formal legal frameworks (Chiarolla, 2012). Such integration not only enhances the legal recognition of farmers' traditional practices but also supports a more resilient and equitable agroecosystem (Coombe, 2011). Central to this reimagining is participatory policy-making, wherein farmer communities are not merely stakeholders but co-creators of legal and policy frameworks (Frison et al., 2011). Mechanisms for inclusive governance—such as community seed banks, farmer-led consultations, and participatory plant breeding initiatives—must be institutionalized to reflect the lived realities and innovations of those working directly with the land (Cleveland & Soleri, 2007; Visser et al., 2005). Moreover, agroecology, as a science, practice, and social movement, provides a vital normative framework through which Farmers' Rights can be re-anchored in environmental justice (Wezel et al., 2009; Altieri, 1995). By aligning agricultural law and policy with agroecological principles, the legal system can move away from extractive models toward regenerative and culturally embedded practices (IPES-Food, 2016). A rights-based agroecological approach foregrounds the agency of farmers as custodians of biodiversity, linking their rights to broader discourses on food sovereignty, climate justice, and intergenerational equity (La Vía Campesina, 2018; De Schutter, 2009). In this way, a reimagined legal landscape transcends the traditional silos of environmental, agricultural, and intellectual property law, offering a more holistic and just framework for sustaining both livelihoods and ecosystems in an era of unprecedented ecological and technological change.

CONCLUSION

The intersection of agricultural intellectual property rights and farmers' rights reveals a complex and often conflicting legal landscape, where the commodification of seeds and biotechnological innovation frequently undermines traditional agricultural practices and community-based seed systems. The prevailing intellectual property frameworks—driven by instruments such as TRIPS, UPOV, and patent regimes—prioritize corporate control and market logic, often at the expense of agrobiodiversity, food sovereignty, and ecological resilience. In contrast, instruments like the ITPGRFA and the UN Declaration on the Rights of Peasants (UNDROP) offer alternative paradigms that emphasize equitable benefit sharing, recognition of traditional knowledge, and participatory governance. The ecological implications of IP-driven monocultures, coupled with the sociopolitical exclusion of indigenous and smallholder voices, underscore the need to reframe farmers' rights not merely as a policy addendum but as a foundational component of sustainable agricultural governance.

To address these challenges, a pluralistic legal approach that integrates statutory intellectual property frameworks with customary law, community seed rights, and agroecological principles. National legal systems should develop inclusive mechanisms for access and benefit-sharing, establish robust safeguards against biopiracy, and enable participatory structures for policymaking. Furthermore, global governance bodies must resolve ambiguities surrounding digital sequence information

(DSI) and synthetic biology to ensure fair use and equitable access in emerging bio-digital spaces.

Future research should deepen inquiry into the lived experiences of farming communities navigating legal complexities, with a focus on gender, indigeneity, and intergenerational knowledge transmission. Comparative legal studies examining hybrid models that harmonize intellectual property protection with collective rights will also be critical. Ultimately, reimagining the legal landscape of farmers' rights requires a shift from extractive legal logics to regenerative, rights-based frameworks that recognize farmers not just as beneficiaries, but as co-creators of agricultural knowledge and biodiversity.

REFERENCES

Alston, J. M., & Venner, R. J. (2002). The effects of the US Plant Variety Protection Act on wheat genetic improvement. *Research Policy*, *31*(4), 527–542. DOI: 10.1016/S0048-7333(01)00123-8

Altieri, M. A. (1995). *Agroecology: The science of sustainable agriculture*. Westview Press. DOI: 10.3362/9781788532310

Altieri, M. A. (2009). Agroecology, small farms, and food sovereignty. *Monthly Review (New York, N.Y.)*, *61*(3), 102–113. DOI: 10.14452/MR-061-03-2009-07_8

Andersen, R. (2008). *Governing Agrobiodiversity: Plant Genetics and Developing Countries*. Ashgate.

Andersen, R., & Winge, T. (2012). *Realising Farmers' Rights to Crop Genetic Resources: Success Stories and Best Practices*. Routledge.

Bagley, M. A. (2020). Digital sequence information: Towards a legal framework to ensure fair and equitable benefit sharing. *International Journal of Law and Technology*, *28*(3), 259–283. DOI: 10.1093/ijlit/eaaa011

Baldwin, A. (2009). Legal ecology: Ecosystem function and the reconfiguration of property. *Environmental Politics*, *18*(4), 588–606.

Boer, B. (2016). *Environmental Law Dimensions of Human Rights*. Oxford University Press.

Borowiak, C. (2004). Farmers' rights: Intellectual property regimes and the struggle over seeds. *Politics & Society*, *32*(4), 511–543. DOI: 10.1177/0032329204269979

Brush, S. B. (2005). *Farmers' Bounty: Locating Crop Diversity in the Contemporary World*. Yale University Press.

Campi, M. (2017). The effect of intellectual property rights on agricultural productivity. *Agricultural Economics*, *48*(3), 327–339. DOI: 10.1111/agec.12337

Campi, M., & Nuvolari, A. (2021). Intellectual property rights and agricultural development: Evidence from a worldwide index of IPRs in agriculture (1961-2018). *The Journal of Development Studies*, *57*(4), 650–668. DOI: 10.1080/00220388.2020.1817395

Capra, F., & Mattei, U. (2015). *The ecology of law: Toward a legal system in tune with nature and community*. Berrett-Koehler Publishers.

CBD. (1992). *Convention on Biological Diversity*. United Nations. https://www.cbd.int

Chaudhary, B. (2013). *Protecting Farmers' Rights in the Global IPR Regime: Challenges and Strategies*. Springer.

Chaudhary, S., Gustafson, D., Mathys, A., & Altieri, M. (2019). A systems approach to assessing agroecology's role in sustainable food systems. *Nature Sustainability*, *2*(5), 393–400.

Chen, J. (2005). The parable of the seeds: Interpreting the Plant Variety Protection Act in furtherance of innovation policy. *The Notre Dame Law Review*, *81*, 105.

Chhetri, N., Easterling, W., Terando, A., & Udry, C. (2022). Intellectual property rights and the global governance of agricultural biotechnology. *World Development*, *151*, 105757. DOI: 10.1016/j.worlddev.2021.105757

Chiarolla, C., & Louafi, S. (2015). Implementing the Multilateral System of Access and Benefit-Sharing: Lessons from the Crop Case. *Biodiversity and Conservation*, *24*(13), 3119–3142.

Clancy, M. S., & Moschini, G. (2017). Intellectual property rights and the ascent of proprietary innovation in agriculture. *Annual Review of Resource Economics*, *9*(1), 53–74. DOI: 10.1146/annurev-resource-100516-053524

Cleveland, D. A., & Soleri, D. (2007). *Farmers, Scientists, and Plant Breeding: Integrating Knowledge and Practice*. CABI.

Coombe, R. J. (2009). The recognition of indigenous peoples' and community traditional knowledge in international law. *St. Thomas Law Review*, *21*(2), 275–298.

Correa, C. M. (2000). *Intellectual Property Rights, the WTO and Developing Countries: The TRIPS Agreement and Policy Options*. Zed Books.

Correa, C. M. (2013). *Plant Variety Protection in Developing Countries: A Tool for Designing a Sui Generis Plant Variety Protection System*. GRAIN.

Cullet, P. (2005). Environmental justice in the use and conservation of biodiversity and water: The role of local communities. *Journal of Environmental Law*, *17*(2), 185–210.

Cullet, P., & Mayet, M. (2011). The sui generis system of seed rights in India: Its implementation and potential. *Journal of Law. Environmental Development*, *7*(2), 248–271.

Curtis, J. M. (2012). Intellectual property rights and international trade: An overview.

De Jonge, B. (2014). *Plant variety protection and farmers' rights: Towards a new balance*. Wageningen Academic Publishers.

De Schutter, O. (2009). *Seed policies and the right to food: Enhancing agrobiodiversity and encouraging innovation*. Report of the Special Rapporteur on the right to food, UN Doc. A/64/170.

Delmer, D. P., Nottenburg, C., Graff, G. D., & Bennett, A. B. (2003). Intellectual property resources for international development in agriculture. *Plant Physiology*, *133*(4), 1666–1670. DOI: 10.1104/pp.103.028993 PMID: 14681529

Dutfield, G. (2008). *Turning plant varieties into intellectual property: The UPOV Convention and its implications for farmers*. Quaker United Nations Office.

Dutfield, G. (2011). *Food, Biological Diversity and Intellectual Property: The Role of the International Union for the Protection of New Varieties of Plants (UPOV)*. Quaker United Nations Office.

FAO. (2009). *International Treaty on Plant Genetic Resources for Food and Agriculture*. Food and Agriculture Organization of the United Nations.

FAO. (2010). *The Second Report on the State of the World's Plant Genetic Resources for Food and Agriculture*.

FAO. (2015). Status of World's Soil resources. https://openknowledge.fao.org/server/api/core/bitstreams/6ec24d75-19bd-4f1f-b1c5-5becf50d0871/content

FAO. (2019). *The state of the world's biodiversity for food and agriculture*. FAO Commission on Genetic Resources for Food and Agriculture.

FAO. (2021). *Agroecology and biodiversity: A pathway to sustainable food systems*. FAO.

Farmers Rights. (2022, December). Why farmers rights matter? https://www.farmersrights.org/what-are-farmers-rights/why-farmers-rights-matter/

Frison, C., López, F., & Esquinas-Alcázar, J. (Eds.). (2011). *Plant genetic resources and food security: Stakeholder perspectives on the International Treaty on Plant Genetic Resources for Food and Agriculture*. Earthscan.

Fuglie, K., & MacDonald, J. M. (2023). Expanded intellectual property protections for crop seeds increase innovation and market power for companies. *Amber Waves: The Economics of Food, Farming, Natural Resources, and Rural America, 2023*.

Ghijsen, H. C. (2002). Property rights on plant varieties: An overview. *Journal of New Seeds*, *4*(1-2), 195–212. DOI: 10.1300/J153v04n01_15

Gibson, J. (2004). *Community resources: Intellectual property, international trade and the protection of traditional knowledge*. Ashgate.

Girsberger, M. (1999). *Biodiversity and the Concept of Farmers' Rights in International Law*. Peter Lang.

GRAIN. (2005). *TRIPS-plus through the back door.*

ETC Group. (2013). *Putting the Cartel before the Horse... and Farm, Seeds, Soil, Peasants, etc.*

Halewood, M., López Noriega, I., & Louafi, S. (Eds.). (2013). *Crop Genetic Resources as a Global Commons: Challenges in International Law and Governance*. Routledge.

Horton, D. (1998). Disciplinary roots and branches of evaluation: Some lessons from agricultural research. *Knowledge in Society*, *10*(4), 31–66. DOI: 10.1007/BF02912498

Howard, P. H. (2009). Visualizing Consolidation in the Global Seed Industry: 1996–2008. *Sustainability (Basel)*, *1*(4), 1266–1287. DOI: 10.3390/su1041266

IPES-Food. (2016). *From uniformity to diversity: A paradigm shifts from industrial agriculture to diversified agroecological systems*. International Panel of Experts on Sustainable Food Systems.

Jarvis, D. I., Hodgkin, T., Sthapit, B. R., Fadda, C., & Lopez-Noriega, I. (2016). A heuristic framework for identifying multiple ways of supporting the conservation and use of traditional crop varieties within the agricultural production system. *Critical Reviews in Plant Sciences*, *30*(1–2), 125–176.

Jasanoff, S., Hurlbut, J. B., & Saha, K. (2021). CRISPR democracy: Gene editing and the need for inclusive deliberation. *Issues in Science and Technology*, *37*(2), 25–32.

Jefferson, O. A., Köllhofer, D., Ehrich, T. H., & Jefferson, R. A. (2020). Transparency tools in gene patenting for informing policy and practice. *Nature Biotechnology*, *38*(4), 405–409. PMID: 24316644

Jewell, C. (2015). Who benefits from IP rights in agricultural innovation. *WIPO Article, August.*

Kiewiet, B. (2005). Plant variety protection in the European Community. *World Patent Information*, *27*(4), 319–327. DOI: 10.1016/j.wpi.2005.07.006

Kingsbury, N. (2009). *Hybrid: The History and Science of Plant Breeding*. University of Chicago Press. DOI: 10.7208/chicago/9780226437057.001.0001

Kloppenburg, J. (2004). *First the Seed: The Political Economy of Plant Biotechnology, 1492–2000* (2nd ed.). University of Wisconsin Press.

Kloppenburg, J. (2010). Impeding Dispossession, Enabling Repossession: Biological Open Source and the Recovery of Seed Sovereignty. *Journal of Agrarian Change, 10*(3), 367–388. DOI: 10.1111/j.1471-0366.2010.00275.x

La Vía Campesina. (2015). *Seeds: Commons or corporate property?*https:// viacampesina.org

Laird, S. A., & Wynberg, R. P. (2018). *A fact-finding and scoping study on digital sequence information on genetic resources in the context of the Convention on Biological Diversity and the Nagoya Protocol*. Secretariat of the Convention on Biological Diversity.

Leskien, D., & Flitner, M. (1997). *Intellectual Property Rights and Plant Genetic Resources: Options for a Sui Generis System*. IPGRI.

Llewelyn, M., & Adcock, M. (2006). *European plant intellectual property*. Bloomsbury Publishing.

Louafi, S. (2020). Governing digital sequence information to ensure fairness in the use of genetic resources. *Frontiers in Plant Science, 11*, 628512. DOI: 10.3389/fpls.2020.628512

Louafi, S. (2022). Digital sequence information and the conservation and sustainable use of biodiversity. *Frontiers in Genetics, 13*, 808459.

Louwaars, N., & De Boef, W. (2012). Integrated Seed Sector Development in Africa: A Conceptual Framework for Creating Coherence between Practices, Programs, and Policies. *Journal of Crop Improvement, 26*(1), 39–59. DOI: 10.1080/15427528.2011.611277

Louwaars, N. P. (2007). *Seeds of Confusion: The Impact of Policies on Seed Systems*. Wageningen University.

Louwaars, N. P., Dons, H., Overwalle, G. V., Raven, H., Arundel, A., Eaton, D., & Nelis, A. (2005). *Breeding Business: The Future of Plant Breeding in the Light of Developments in Patent Rights and Plant Breeders' Rights*. Centre for Genetic Resources.

Migliori, I. K. (2021). The current status of plant-related IP Rights in Brazil: How to reap the fruits of the protection culture in the agribusiness field. *Stockholm Intellectual Property Law Review*, (1), 48–55.

Moeller, N. I., & Stannard, C. (2013). Identifying benefit flows. Studies on the Potential Monetary and Non-Monetary Benefits Arising from the International Treaty on Plant Genetic Resources for Food and Agriculture, FAO-Treaty Secretariat.

Montoliu, L. (2019). The CRISPR revolution: Impact on genomics, biotechnology and society. *EMBO Reports*, *20*(9), e48595. DOI: 10.15252/embr.201948595

Mooney, P. R. (2011). The law of the seed: Another development and plant genetic resources. *Development Dialogue*, *1*(1), 1–75.

Nijar, G. S. (2011). *The Nagoya Protocol on Access and Benefit Sharing: An Analysis*. Third World Network.

Oguamanam, C. (2014). Farmers' rights and the intellectual property dynamic in agriculture. *Sage Handbook of Intellectual Property*, 238-257.

Pauchard, N., & Godt, C. (2021). Legal challenges of digital sequence information: Implications for access and benefit-sharing. *Journal of Environmental Law*, *33*(1), 1–25. DOI: 10.1093/jel/eqa

Peschard, K. (2017). Seed wars and farmers' rights: Comparative perspectives from Brazil and India. *The Journal of Peasant Studies*, *44*(1), 144–168. DOI: 10.1080/03066150.2016.1191471

Peschard, K., & Randeria, S. (2020). 'Keeping seeds in our hands': Legal pluralism and seed governance in the Global South. *Third World Quarterly*, *41*(6), 1012–1030.

Pimbert, M. (2009). *Towards food sovereignty: Reclaiming autonomous food systems*. IIED.

Posey, D. A., & Dutfield, G. (1996). *Beyond Intellectual Property: Toward Traditional Resource Rights for Indigenous Peoples and Local Communities*. IDRC.

Prakash, A. (2017). The Politics of Seed Laws in India: A Historical Institutional Analysis. *The Journal of Peasant Studies*, *44*(1), 1–22.

Prasad, P. V., Bhatnagar, N., Bhandari, V., Jacob, G., Narayan, K., Echeverría, R., Beintema, N., Farah Cox, P., & Compton, J. (2023). Patterns of investment in agricultural research and innovation for the Global South, with a focus on sustainable agricultural intensification. *Frontiers in Sustainable Food Systems*, *7*, 1108949. DOI: 10.3389/fsufs.2023.1108949

Prip, C. (2024). Farmers' Rights in the Plant Treaty: Interrelations and recent interactions with other international regimes and processes. *Development Studies Research*, *11*(1), 2357095. Advance online publication. DOI: 10.1080/21665095.2024.2357095

RAFI. (1998). *Patents, plants and plunder*. Rural Advancement Foundation International.

Raj, K. (2007). *Relocating Modern Science: Circulation and the Construction of Knowledge in South Asia and Europe, 1650–1900*. Palgrave Macmillan. DOI: 10.1057/9780230625310

Ramanna, A. (2006). India's Plant Variety and Farmers' Rights Legislation: Potential Benefits and Harms. *The Journal of World Intellectual Property*, *9*(2), 189–210.

Rout, S. K. (2012). Protection of Plant Varieties and Farmers' Rights in India: A Critical Appraisal. *Journal of Intellectual Property Rights*, *17*, 478–489.

Sahai, S. (2003). India's Plant Variety Protection and Farmers' Rights Act, 2001. *Current Science*, *84*(3), 388–393.

Sahai, S. (2003). India's Plant Variety Protection and Farmers' Rights Act, 2001. *Current Science*, *84*(3), 411–417.

Scholz, A. H., Huft, M., & Tvedt, M. W. (2020). Bio-digital prospecting and the dilemma of open science. *Nature Biotechnology*, *38*(4), 389–391. DOI: 10.1038/s41587-020-0482-9

Semon, S. F. A. (2003, September). Developments in breeding and techniques within the community plant variety rights fruit sector. In *XI Eucarpia Symposium on Fruit Breeding and Genetics 663* (pp. 707-712).

Shiva, V. (1997). *Biopiracy: The Plunder of Nature and Knowledge*. South End Press.

Shiva, V. (2000). *Stolen harvest: The hijacking of the global food supply*. South End Press.

Shiva, V. (2016). *Who Really Feeds the World? The Failures of Agribusiness and the Promise of Agroecology*. North Atlantic Books.

Stringer, L. C., Fraser, E. D., Harris, D., Lyon, C., Pereira, L., Ward, C. F., & Simelton, E. (2020). Adaptation and development pathways for different types of farmers. *Environmental Science & Policy*, *104*, 174–189. DOI: 10.1016/j.envsci.2019.10.007

Tonapi, V. A. (2008). PPV and FR Act and National Seed Policy. *Sorghum Improvement in the New Millennium. Patancheru 502 324, Andhra Pradesh, India: International Crops Research Institute for the Semi-Arid Tropics. 340 pp, 48*, 232.

United Nations. (2018). *United Nations Declaration on the Rights of Peasants and Other People Working in Rural Areas (UNDROP)*. https://www.ohchr.org

Velasco, A. C. (2020). The CJEU reaches a controversial compromise on the scope of protection of plant variety rights. *Journal of Intellectual Property Law & Practice*, *15*(12), 976–983. DOI: 10.1093/jiplp/jpaa159

Venkatesan, R. (2018). TRIPS and Plant Variety Protection in India: Complicating the Globalisation Debate. *Indian J. Int'l Econ. L.*, *9*, 43.

Verma, D. D. P. (2010). Review of Article 27.3 (B) Under TRIPs Agreement: A Critical Analysis. *Available atSSRN* 1587584. DOI: 10.2139/ssrn.1587584

Visser, B., Eaton, D., Louwaars, N., & van Dusen, M. E. (2005). *Transaction costs of germplasm exchange under bilateral agreements*. Global Forum on Agricultural Research.

Wezel, A., Bellon, S., Doré, T., Francis, C., Vallod, D., & David, C. (2009). Agroecology as a science, a movement and a practice. *Sustainable Agriculture*, *29*(4), 503–515.

Winge, T., Andersen, R., & Ramanna-Pathak, A. (2013). Combining Farmers' Rights and plant variety protection in Indian law. In *Realising Farmers' Rights to Crop Genetic Resources* (pp. 54-61). Routledge.

World Bank Group. (2025). Agriculture and Food. https://www.worldbank.org/en/topic/agriculture/overview

WTO. (1994). *The Agreement on Trade-Related Aspects of Intellectual Property Rights (TRIPS).*

Wynberg, R., & Chennells, R. (2009). Green Governance: An Institutional Analysis of ABS and Traditional Knowledge. In Wynberg, R., Schroeder, D., & Chennells, R. (Eds.), *Indigenous Peoples, Consent and Benefit Sharing* (pp. 189–206). Springer. DOI: 10.1007/978-90-481-3123-5

KEY TERMS AND DEFINITIONS

Farmers' Rights: A set of entitlements recognizing the contributions of indigenous and local communities to the conservation and development of plant genetic resources, encompassing rights to save, use, exchange, and sell farm-saved seed, and to participate in decision-making and benefit-sharing processes (ITPGRFA, Art. 9).

Agricultural Intellectual Property (Ag-IP): Legal rights, including patents, plant variety protections, and trade secrets, that confer exclusive control over agricultural innovations such as genetically modified seeds, crop biotechnologies, and breeding techniques.

Agrobiodiversity: The variety and variability of animals, plants, and microorganisms used directly or indirectly for food and agriculture, sustained primarily by traditional agricultural communities through diverse and dynamic farming systems.

Access and Benefit Sharing (ABS): A legal mechanism derived from the Convention on Biological Diversity (CBD) and the Nagoya Protocol, aimed at ensuring fair and equitable sharing of benefits arising from the utilization of genetic resources, particularly with provider countries and indigenous communities.

UPOV Convention: The International Convention for the Protection of New Varieties of Plants (UPOV), which provides a framework for granting exclusive breeders' rights, is often criticized for limiting farmers' traditional seed practices and privileged commercial breeders.

Customary Law: Community-based norms and practices that are informally observed and transmitted across generations, governing access to and use of natural resources, including seeds, outside of formal state-based legal systems.

Seed Sovereignty: The right of farmers to breed, save, exchange, and sell seeds freely, grounded in principles of local control over genetic resources, cultural preservation, and resistance to corporate monopolization of seed systems.

Digital Sequence Information (DSI): Genetic data derived from the sequencing of biological resources, increasingly central to agricultural biotechnology and raising contentious legal questions around ownership, benefit-sharing, and open access in global governance regimes.

Legal Pluralism: The coexistence and interaction of multiple legal systems—formal, customary, religious, and international—within a single socio-political context, relevant to the governance of farmers' rights and plant genetic resources.

Chapter 6
Harnessing AI for Environmental Hazard Mitigation and Resilience:
Pathways to a Safer Future

Ananya Pandey
https://orcid.org/0000-0002-2419-6314
Christ University, Bangalore, India

ABSTRACT

This chapter explores the transformative potential of artificial intelligence (AI) in mitigating environmental hazards and enhancing resilience to climate-related challenges. AI technologies are revolutionizing environmental monitoring, risk assessment, and disaster response through advanced data analytics, machine learning, and predictive modeling. By leveraging AI, stakeholders can predict and mitigate the impacts of natural disasters, improve resource management, and foster resilient infrastructures. The chapter examines the various applications of AI in hazard prediction, such as flood forecasting, wildfire management, and storm tracking, alongside its role in real-time monitoring and adaptive response systems. Furthermore, it highlights the ethical considerations, challenges, and limitations of deploying AI in environmental management. Through case studies and emerging practices, this chapter outlines pathways for harnessing AI to safeguard communities and ecosystems, ensuring a more resilient and sustainable future.

DOI: 10.4018/979-8-3373-3206-2.ch006

INTRODUCTION

The increasing frequency and intensity of environmental hazards pose one of the greatest challenges to humanity in the 21st century. Climate change continues to amplify the severity of natural disasters such as hurricanes, wildfires, droughts, and floods, while industrialization and urbanization contribute to technological hazards including chemical spills and infrastructure failures (IPCC, 2023). These hazards not only threaten ecosystems but also endanger human lives, displace communities, and destabilize economies. In an interconnected global society, localized environmental catastrophes can have far-reaching impacts, highlighting the urgent need for proactive, resilient strategies to mitigate these risks.

The urgency for resilience has never been more pronounced. Resilience, in this context, refers to the ability of individuals, communities, institutions, and systems to anticipate, prepare for, respond to, and recover from environmental hazards effectively (Cutter et al., 2008). Traditional approaches, relying heavily on historical data and reactive policies, have proven insufficient against the dynamic and complex nature of modern environmental threats. The unpredictability of climate systems, the cascading effects of technological failures, and the rapidity of certain natural disasters call for a paradigm shift toward predictive, adaptive, and integrated risk management strategies. Building resilience today demands not only stronger infrastructure and preparedness measures but also innovative tools capable of analyzing vast amounts of environmental data, forecasting potential threats, and facilitating timely interventions.

Artificial Intelligence (AI) has emerged as a transformative tool in this endeavor. By leveraging machine learning algorithms, neural networks, natural language processing, and robotics, AI can revolutionize how we understand, predict, and respond to environmental hazards (Rolnick et al., 2022). AI systems can process real-time data from satellites, sensors, drones, and social media to provide early warning signals, map risk landscapes, and even orchestrate emergency responses. Furthermore, AI can aid in post-disaster recovery by assessing damage, optimizing resource allocation, and supporting decision-makers in rebuilding more resilient communities. The integration of AI into environmental hazard mitigation represents a shift from reactive to proactive management — a vital progression as the window for climate action narrows.

Beyond its technical capabilities, AI introduces opportunities for systemic change. Predictive models powered by AI can simulate various disaster scenarios, allowing governments, corporations, and civil society to prepare for a range of contingencies. AI-enabled digital twins of cities or ecosystems offer virtual environments for testing resilience strategies without risking real-world harm (Batty, 2018). Moreover, AI-driven platforms can democratize access to risk information, empowering vulnerable communities with timely, localized knowledge that enhances their adaptive capacity (Vinuesa et al., 2020). Thus, AI is not merely a technological advancement; it is a catalyst for reshaping environmental governance, promoting inclusivity, and advancing sustainability.

However, the use of AI in hazard mitigation is not without its challenges. Ethical concerns surrounding data privacy, algorithmic bias, and technological disparities must be carefully addressed to ensure that AI solutions benefit all sectors of society equitably (Floridi et al., 2018). Additionally, the resilience of AI systems themselves must be considered, particularly in scenarios where critical infrastructure relies on AI-driven decision-making. Designing transparent, explainable, and accountable AI systems will be crucial for building public trust and ensuring robust, sustainable applications in disaster management.

Given these opportunities and challenges, this chapter aims to explore how AI can be harnessed to mitigate environmental hazards and build resilient systems. It will begin by defining key concepts — environmental hazards, resilience, and the technological foundations of AI. Subsequently, it will examine current applications of AI in hazard prediction, emergency response, and resilience building. Real-world case studies will illustrate successful integrations of AI technologies in disaster risk reduction. The chapter will then address the limitations and ethical considerations associated with AI in this domain. Finally, it will propose future pathways, emphasizing the importance of interdisciplinary collaboration, policy innovation, and inclusive technology development.

By systematically analyzing the intersection of AI and environmental hazard management, this chapter seeks to contribute to a growing body of scholarship and practice that envisions a safer, more resilient future. It invites readers — from policymakers and technologists to community leaders and academics — to consider AI not just as a tool for hazard mitigation, but as an enabler of systemic transformation toward sustainability, equity, and human security. In doing so, it highlights that while environmental hazards may be an inevitable part of our shared future, the extent of their impact — and our ability to recover — remains firmly within our collective control.

UNDERSTANDING ENVIRONMENT HAZARDS AND RESILIENCE

Environmental hazards are defined as naturally occurring or human-induced phenomena that have the potential to cause harm to human health, property, ecosystems, or the built environment (Smith, 2013). They encompass a wide range of threats, including but not limited to natural disasters such as earthquakes, hurricanes, floods, and wildfires, as well as technological hazards like industrial accidents, oil spills, and chemical exposures. In recent decades, the distinction between natural and anthropogenic hazards has become increasingly blurred, as human activities exacerbate the frequency and intensity of natural events through processes like climate change, deforestation, and urbanization (UNDRR, 2019). Understanding the nature, causes, and impacts of environmental hazards is foundational to developing effective mitigation and resilience strategies.

Natural hazards are typically categorized into geophysical (earthquakes, tsunamis, volcanic eruptions), meteorological (storms, extreme temperatures), hydrological (floods, landslides), climatological (droughts, wildfires), and biological (epidemics, infestations) phenomena (Alexander, 2020). These events, although often occurring without human intervention, can be intensified by human-driven environmental degradation. For instance, deforestation can increase the likelihood of landslides, and global warming is amplifying the frequency of heatwaves and severe storms. Technological or anthropogenic hazards, on the other hand, arise from human activities — industrial explosions, toxic waste spills, nuclear accidents — and often involve complex interactions between technological systems and human behavior (Perrow, 2011).

A crucial dimension of hazard assessment is vulnerability, which refers to the susceptibility of a community or system to harm (Cutter, 2006). Vulnerability is influenced by various factors, including socioeconomic status, geographic location, governance capacity, infrastructure quality, and access to resources. Consequently, similar hazard events can have vastly different impacts across regions and populations. For example, a hurricane striking a well-prepared, economically stable city may result in limited casualties, while the same event in a poorer, less resilient community can be devastating.

The concept of resilience offers a proactive and hopeful framework for addressing environmental hazards. In essence, resilience is the ability of a system — whether ecological, social, economic, or infrastructural — to absorb disturbances, adapt to changing conditions, and rapidly recover from disruptions (Walker et al., 2004). In disaster management, resilience shifts the focus from solely preventing hazards — an often-impossible task — to enhancing a system's capacity to withstand and rebound from adverse events.

Resilience has multiple dimensions: **engineering resilience**, emphasizing the ability to return to a previous state after disruption; **ecological resilience**, highlighting the capacity for systems to absorb shocks and maintain essential functions; and **social resilience**, focusing on communities' abilities to organize, adapt, and transform in response to crises (Folke et al., 2010). These perspectives are not mutually exclusive; effective hazard mitigation strategies often integrate elements of all three.

Building resilience involves both structural and non-structural measures. Structural measures include physical interventions like levees, seawalls, earthquake-resistant buildings, and smart energy grids. Non-structural measures involve policies, education, community engagement, and governance reforms that enhance adaptive capacities (Tierney, 2014). Importantly, resilience is not solely about bouncing back to a pre-disaster state but can also involve "bouncing forward" — leveraging crises as opportunities for transformation and systemic improvement.

The integration of resilience thinking into disaster risk management has been codified in global frameworks such as the Sendai Framework for Disaster Risk Reduction 2015–2030. This framework emphasizes understanding disaster risk, strengthening disaster risk governance, investing in resilience, and enhancing disaster preparedness for effective response (UNDRR, 2015). Central to these efforts is the recognition that resilience must be community-centered, inclusive, and informed by scientific evidence, traditional knowledge, and technological innovation.

Yet, significant challenges remain in operationalizing resilience across different contexts. Measuring resilience quantitatively remains difficult due to its complex, multidimensional nature (Cutter et al., 2010). Additionally, resilience-building efforts often require long-term investments, cross-sector collaboration, and political will, which can be difficult to sustain, especially in resource-constrained settings. There is also the risk of "resilience inequities," where wealthier communities are better able to invest in resilience measures, while vulnerable populations remain disproportionately exposed to hazards (Matin et al., 2018).

In this regard, the advent of new technologies, particularly artificial intelligence (AI), offers promising tools for enhancing both our understanding of environmental hazards and our capacity to build resilience. AI's ability to process large, complex datasets enables more accurate hazard mapping, risk assessment, and early warning systems (Rolnick et al., 2022). Machine learning models can identify patterns in historical disaster data to predict future events, while natural language processing can analyze social media feeds in real time to track emerging crises. Moreover, AI-driven decision support systems can help planners design infrastructure and urban systems that are better able to adapt to changing environmental conditions.

AI technologies also present opportunities for engaging communities in resilience-building efforts. Citizen science platforms powered by AI can aggregate local knowledge and observations, creating more comprehensive and context-sensitive risk assessments (Vinuesa et al., 2020). Furthermore, AI can facilitate more equitable resilience by making risk information more accessible, thereby empowering marginalized communities to participate actively in disaster preparedness and recovery processes.

However, just as resilience must be holistic and inclusive, so too must the development and deployment of AI in this field. Ethical considerations regarding data privacy, algorithmic bias, and technological accessibility must be at the forefront to ensure that AI strengthens resilience equitably across all segments of society (Floridi et al., 2018). Without careful design and governance, AI applications risk reinforcing existing inequalities and creating new vulnerabilities.

In sum, environmental hazards represent a complex and evolving threat landscape that demands innovative, systemic responses. Understanding the multifaceted nature of hazards and the dynamics of resilience provides a critical foundation for exploring how emerging technologies, particularly AI, can be harnessed to create safer, more adaptive societies. This chapter builds on these foundational concepts to examine the specific ways in which AI is already transforming hazard mitigation and resilience efforts — and how it can be further leveraged to secure a safer future.

THE PROMISE OF ARTIFICIAL INTELLIGENCE

Artificial Intelligence (AI) represents a paradigm shift in our ability to confront and mitigate environmental hazards. With the capability to process enormous datasets, recognize complex patterns, and deliver predictive insights, AI technologies offer unprecedented opportunities to enhance the speed, accuracy, and efficiency of disaster risk management (Russell & Norvig, 2021). As climate change, urbanization, and globalization intensify environmental vulnerabilities, AI emerges not merely as a supplementary tool but as a central actor in reimagining resilience and proactive hazard mitigation.

At its core, AI involves the creation of machines or systems capable of performing tasks that typically require human intelligence, such as perception, reasoning, learning, and decision-making (Goodfellow et al., 2016). These capabilities are driven by techniques such as machine learning, deep learning, natural language processing (NLP), and computer vision. When applied to environmental hazard management, these techniques can deliver real-time situational awareness, predictive analytics, automated decision-support systems, and autonomous operational responses — all of which are critical in environments where every second counts.

One of AI's most transformative promises lies in early warning systems. Predictive models powered by machine learning can analyze historical hazard data, meteorological patterns, satellite imagery, and ground sensor inputs to forecast potential disasters with greater accuracy and lead time than traditional systems (Liu et al., 2022). For instance, AI-enhanced flood forecasting models integrate real-time rainfall data, river gauge measurements, and terrain information to predict flood events with high spatial and temporal resolution. This not only enables authorities to issue timely evacuation orders but also allows critical infrastructure operators to preemptively safeguard essential services.

Beyond early warning, AI significantly enhances disaster preparedness and risk assessment. Through high-resolution risk mapping, AI models can identify vulnerable zones based on variables such as elevation, land use, demographic data, and infrastructure resilience (Chollet, 2021). For example, convolutional neural networks (CNNs) — a form of deep learning — have been used to analyze satellite imagery and assess wildfire risks by identifying patterns of vegetation dryness and human encroachment. Such analyses help urban planners, insurers, and emergency managers prioritize resource allocation, fortify at-risk communities, and inform land-use policies.

AI also plays a pivotal role in disaster response and recovery. During an ongoing crisis, AI-powered drones equipped with computer vision algorithms can rapidly survey affected areas, assess damages, and identify people in need of rescue (Gao et al., 2022). NLP algorithms can sift through vast volumes of social media posts, emergency calls, and news reports to detect emerging hotspots and direct first responders more efficiently. In the aftermath of disasters, AI can optimize the logistics of humanitarian aid distribution, ensuring that food, water, medicine, and shelter reach the hardest-hit populations without unnecessary delays or duplication.

An equally promising but less visible application of AI is in scenario modeling and simulation. Digital twins — virtual replicas of physical systems enhanced by real-time data and AI analytics — allow decision-makers to simulate various disaster scenarios and stress-test resilience strategies (Batty, 2018). For example, a digital twin of a coastal city can be used to model the impacts of different storm surge intensities, allowing planners to evaluate the effectiveness of seawalls, evacuation routes, and emergency shelter networks under varying conditions. These simulations inform not only emergency responses but also long-term investments in infrastructure and urban design.

Importantly, AI has the potential to democratize hazard mitigation by empowering local communities. Citizen science platforms, mobile apps, and community-driven data initiatives can be enhanced by AI to create participatory risk assessments and real-time feedback loops (Vinuesa et al., 2020). Such tools allow marginalized communities, often the most vulnerable to environmental hazards, to contribute directly to data collection, hazard detection, and decision-making processes. By broadening participation, AI can help address some of the equity gaps historically entrenched in disaster management practices.

Despite these transformative potentials, it is crucial to recognize that the promise of AI is not automatic. Its efficacy in environmental hazard mitigation depends heavily on the quality, representativeness, and accessibility of the data on which it is trained. Data gaps, particularly in low-resource settings, can lead to blind spots in AI models, reducing their predictive power and reinforcing existing inequalities (Beck et al., 2020). Additionally, AI models are often criticized for their "black box" nature — their internal decision-making processes are opaque, making it difficult for users to trust or understand their outputs (Burrell, 2016).

Ethical considerations further complicate the deployment of AI in disaster contexts. Privacy concerns arise when AI systems collect and analyze personal data from sensors, drones, mobile devices, and social media during emergencies. Moreover, algorithmic biases — reflecting historical and systemic inequalities embedded in training data — can skew risk assessments and disaster responses in ways that disadvantage already marginalized groups (O'Neil, 2016). Ensuring transparency, accountability, and fairness in AI design and application is therefore not optional; it is essential for building technologies that genuinely enhance resilience.

Another significant challenge is ensuring the resilience of AI systems themselves. AI applications rely on digital infrastructures — servers, communication networks, cloud services — that are vulnerable to disruption during disasters. Power outages, cyberattacks, and physical damage to IT facilities can render AI systems inoperable precisely when they are most needed. Building redundancies, offline capabilities, and robust cybersecurity measures are critical to making AI systems as resilient as the communities they aim to serve (Subramanian et al., 2021).

Moreover, interdisciplinary collaboration is vital to realizing AI's promise in this field. AI developers must work closely with climate scientists, emergency managers, urban planners, policymakers, ethicists, and affected communities to ensure that AI solutions are grounded in real-world needs, scientific rigor, and social justice principles (Rolnick et al., 2022). No single discipline holds all the knowledge necessary to address the complex, multi-dimensional nature of environmental hazards.

AI IN ACTION FOR ENVIRONMENTAL HAZARD MITIGATION

The real-world application of artificial intelligence (AI) for environmental hazard mitigation is no longer hypothetical. Across the globe, diverse AI-driven projects illustrate how technology can revolutionize the way we anticipate, manage, and recover from disasters. These case studies not only demonstrate the practical capabilities of AI but also highlight the challenges and lessons learned in deploying these technologies at scale.

Predicting Wildfires with AI: The Firecast Model

In recent years, wildfires have grown in intensity and frequency, particularly in areas such as California, Australia, and the Mediterranean. Traditional prediction models, which rely heavily on weather conditions and historical fire data, often struggle with rapidly changing environmental variables. Recognizing this gap, the World Resources Institute (WRI) partnered with NASA and tech firms to develop Firecast, an AI-powered early warning system (Cheng et al., 2022).

Firecast employs deep learning models trained on satellite imagery, vegetation indices, meteorological data, and historical fire patterns. Through convolutional neural networks (CNNs), the system identifies regions with heightened fire risk days or even weeks in advance. These forecasts allow local governments and emergency services to implement preemptive measures such as controlled burns, public warnings, and strategic resource deployment.

However, deploying Firecast has underscored several challenges. The model's predictive accuracy depends heavily on continuous satellite data feeds, which can be interrupted by cloud cover or equipment failures. Moreover, there is an ongoing need to update training datasets to reflect the changing behavior of wildfires driven by climate change. Despite these challenges, Firecast marks a significant advancement in proactive wildfire mitigation.

Flood Prediction in India: IBM's Operation Blue Sky

Flooding is the most frequent natural disaster worldwide, and countries like India, with densely populated floodplains, face significant risks. In 2018, IBM Research launched Operation Blue Sky, a project to create hyperlocal flood forecasting models using AI in collaboration with Indian authorities (Patel et al., 2019).

The system integrates weather forecasts, river level data, topography, soil moisture indices, and urban drainage models into a machine learning framework. By combining decision trees and ensemble learning techniques, the system predicts when and where flooding is likely to occur, down to specific neighborhoods. Importantly, the forecasts are delivered via mobile alerts to local communities, offering crucial hours of advanced warning.

Operation Blue Sky has been credited with reducing casualties and economic losses in pilot areas. Nevertheless, it faced obstacles such as inconsistent data quality, bureaucratic resistance to adopting AI-generated predictions over traditional methods, and the challenge of effectively reaching marginalized populations without smartphones or internet access. These insights emphasize the importance of inclusive design and robust stakeholder engagement in AI projects.

Earthquake Damage Assessment: Facebook's Machine Learning Maps

While earthquakes are difficult to predict, rapid damage assessment post-disaster is critical for effective response. Facebook's Data for Good initiative developed **machine learning maps** that provide high-resolution population density data, which, when combined with seismic impact models, help estimate the number of people affected after an earthquake (Maas et al., 2020).

Following the 2019 Albanian earthquake, Facebook's AI-generated maps enabled relief agencies to prioritize search and rescue operations in the most densely populated and heavily affected areas. The models used AI-powered object detection algorithms to differentiate between buildings, open spaces, and infrastructure across millions of satellite images.

The case highlights the role of public-private partnerships in enhancing disaster resilience. Yet, it also points to ethical concerns about data privacy and the dependence of humanitarian efforts on proprietary corporate data. As AI continues to permeate disaster management, ensuring that data use aligns with ethical standards and humanitarian principles becomes increasingly critical.

Tsunami Early Warning: Japan's AI-Enhanced Systems

Japan, located along the seismically active Pacific Ring of Fire, has long invested heavily in disaster preparedness. After the devastating 2011 Tōhoku earthquake and tsunami, Japanese agencies collaborated with researchers to integrate AI into tsunami early warning systems (Mori et al., 2019).

AI models now process seismic data, ocean buoy readings, and undersea pressure sensors in real-time to predict tsunami wave height, arrival time, and potential inundation zones within minutes of an earthquake. Machine learning algorithms also correct initial rough estimates as new data flows in, improving accuracy as the situation evolves.

This AI-enhanced system dramatically reduces false alarms, which had historically plagued Japan's tsunami warnings, and enables more targeted evacuations. However, real-world drills and public education are crucial in ensuring that AI-generated warnings lead to appropriate human action. Technology alone is insufficient without societal readiness to act on the information provided.

Monitoring Air Pollution: Google's Project Air View

Environmental hazards are not limited to natural disasters; industrial accidents and chronic pollution also pose significant threats. Google's Project Air View uses AI-powered sensors mounted on Street View cars to map air quality at a hyperlocal level across urban environments (Apte et al., 2017). The data, analyzed through machine learning algorithms, identifies pollution hotspots, tracks trends over time, and suggests areas where policy interventions or public health measures are most needed. In Oakland, California, the project revealed startling air quality disparities between neighborhoods separated by only a few blocks, providing empirical support for targeted environmental justice initiatives. While innovative, Project Air View also highlights challenges such as the need for standardized pollution monitoring protocols and the risk of creating surveillance anxieties among communities. Transparent communication and community involvement are essential to ensure that such technologies serve the public interest.

ETHICAL, LEGAL AND GOVERNANCE CONSIDERATIONS FOR AI IN ENVIRONMENTAL RESILIENCE

While artificial intelligence (AI) offers transformative potential for mitigating environmental hazards and enhancing resilience, it simultaneously raises complex ethical, legal, and governance challenges. The high stakes involved — including human safety, ecological integrity, and societal equity — make it imperative to scrutinize the deployment of AI technologies in this domain. Addressing these considerations thoughtfully is essential to ensuring that AI solutions do not inadvertently create new risks or exacerbate existing vulnerabilities.

Ethical Considerations: Fairness, Accountability, and Transparency

The ethical use of AI for environmental resilience demands a commitment to fairness, accountability, and transparency. A major concern is the potential for algorithmic bias, where AI systems trained on incomplete or skewed data could disproportionately disadvantage marginalized communities (Mehrabi et al., 2021). For instance, flood prediction models that rely on satellite data may be less accurate in informal settlements lacking formal infrastructure, potentially leading to insufficient warnings or resource allocation.

Transparency, often referred to as the "right to explanation," is another cornerstone ethical principle (Floridi et al., 2018). Stakeholders must be able to understand how AI systems reach their conclusions, particularly in life-or-death situations such as evacuation orders or emergency response prioritization. Opaque AI decision-making processes, often labeled "black boxes," can erode public trust and hinder effective action.

Accountability mechanisms must also be clearly defined. When AI systems make erroneous predictions or recommendations, it must be clear who is responsible — the developers, the operators, or the decision-makers relying on AI outputs. Without clear lines of accountability, the risk of "moral crumple zones" — where blame is shifted onto technology rather than human overseers — increases (Elish, 2019).

Data Governance and Privacy Issues

The effective operation of AI systems for environmental hazard mitigation depends on large-scale data collection, including sensitive information such as geolocation, property details, and sometimes even personal identifiers. This raises critical data privacy and governance issues.

Robust frameworks are needed to govern how data is collected, stored, shared, and utilized (Zwitter & Gstrein, 2020). Consent mechanisms must be embedded wherever personal data is involved, particularly in community-level disaster response platforms that might track individual movements during evacuations. Moreover, data sharing across agencies and borders must balance the need for comprehensive environmental monitoring with the protection of individual rights.

Anonymization and differential privacy techniques can help mitigate risks, but they must be implemented carefully to avoid de-anonymization through dataset correlation (Narayanan & Shmatikov, 2008). In the absence of clear regulatory guidelines tailored to disaster management contexts, developers and policymakers must proactively adopt best practices from broader data ethics discussions.

Legal Frameworks and Liability Challenges

The legal landscape for AI in environmental resilience is still nascent. Existing disaster management laws often do not account for the complexities introduced by autonomous decision-making systems. Key legal challenges include questions of liability, compliance, and jurisdiction.

For instance, if an AI-powered flood warning system fails to predict an inundation correctly, leading to loss of life or property, determining liability can be difficult. Traditional legal doctrines such as negligence, strict liability, or product liability might be applied, but they were not designed with autonomous, learning systems in mind (Calo, 2016). There is a growing argument for creating specialized legal doctrines or regulatory frameworks that address the unique characteristics of AI.

Furthermore, compliance with environmental and disaster regulations becomes complex when AI models operate across jurisdictions with differing legal requirements. International cooperation will likely be essential in setting baseline standards for AI use in transboundary environmental risk management, such as flood prediction across river basins shared by multiple countries.

Governance Frameworks for Ethical AI Deployment

Effective governance mechanisms are crucial to ensure that AI deployment for environmental resilience adheres to ethical and legal standards. Several frameworks have been proposed by international organizations, academic institutions, and professional associations.

For example, UNESCO's Recommendation on the Ethics of Artificial Intelligence (2021) outlines principles such as human oversight, sustainability, and inclusiveness, which are particularly relevant in the context of disaster risk reduction. The OECD's AI Principles also emphasize robustness, safety, and accountability (OECD, 2019).

At the operational level, AI Ethics Impact Assessments (AIEIAs) can be used to evaluate projects before deployment, similar to Environmental Impact Assessments (EIA). AIEIAs involve systematically assessing potential ethical risks, biases, and social impacts of AI systems, enabling mitigation strategies to be integrated early in the project lifecycle (Fjeld et al., 2020).

Furthermore, multi-stakeholder governance models — involving governments, civil society, the private sector, and affected communities — can promote more inclusive and equitable AI development. By incorporating diverse perspectives, these models help ensure that AI systems serve broad societal interests rather than narrow commercial or political goals.

Trust and Public Engagement

Ultimately, the success of AI in environmental resilience hinges on public trust. Even the most technically advanced systems will fail if the communities they intend to protect do not trust them or refuse to act on their warnings. Building trust requires sustained public engagement, education, and co-creation of AI solutions.

Public education campaigns can demystify AI and explain its role in disaster mitigation, countering misinformation and fear. Participatory design processes, where community members help shape AI applications, not only enhance the relevance and usability of these systems but also empower communities, fostering a sense of ownership and responsibility (Winfield & Jirotka, 2018).

Moreover, transparent communication about the limitations of AI — including uncertainties in predictions and the potential for errors — is vital. Overstating AI's capabilities can create a false sense of security, while underplaying them can breed unnecessary skepticism. A balanced narrative, grounded in realism and humility, is essential for ethical AI governance.

CHALLENGES AND LIMITATIONS

Despite the tremendous promise of artificial intelligence (AI) in environmental hazard mitigation, its application is fraught with significant challenges and limitations. These obstacles span technical, operational, ethical, and societal domains. Understanding these issues is critical to developing more robust, equitable, and effective AI-based solutions for enhancing environmental resilience.

Data Quality and Availability

AI systems are only as effective as the data they are trained on. In the domain of environmental hazard mitigation, data quality and availability remain persistent challenges. Many regions, particularly in developing countries, suffer from inadequate sensor networks, poor historical records, and fragmented datasets (Baker et al., 2019). Even in data-rich areas, environmental data can be noisy, incomplete, or biased — factors that impair model accuracy and reliability.

Moreover, extreme events like earthquakes, wildfires, and floods are inherently rare and unpredictable, making it difficult to gather sufficient high-quality data for training predictive models (Kreibich et al., 2017). Synthetic data generation and transfer learning offer potential solutions, but they introduce their own limitations and risks, including the possibility of embedding unrealistic assumptions into models.

Model Complexity, Interpretability, and Trust

Many AI models used in environmental applications, such as deep neural networks, are highly complex and lack interpretability. This "black box" nature makes it difficult for stakeholders to understand how predictions are generated, undermining trust and impeding adoption (Rudin, 2019).

In disaster risk management, where decisions often involve human lives, the need for explainability is paramount. Emergency responders, policymakers, and affected communities must be able to comprehend the basis of AI recommendations to act effectively and confidently. Current advances in explainable AI (XAI) offer partial solutions but are not yet sufficient for all contexts (Samek et al., 2017).

Thus, the trade-off between model accuracy and interpretability remains a fundamental tension in the deployment of AI for environmental resilience.

Generalizability and Transferability

AI models trained for hazard prediction or response in one geographic or socioeconomic context may not generalize well to others. Differences in terrain, infrastructure, governance systems, and population behavior can significantly impact the performance of AI models (Boudreau et al., 2021).

For example, an AI model optimized for predicting wildfire spread in California's Mediterranean climate may perform poorly in Australia's different vegetation and climate regimes. Similarly, flood models developed for urbanized, high-income cities may fail in informal settlements with distinct drainage and construction patterns.

Addressing this limitation requires the development of flexible, adaptive models that can be recalibrated for diverse local contexts, often necessitating significant additional data collection and customization efforts.

Infrastructure and Technical Barriers

Deploying AI systems for environmental hazard mitigation often requires substantial infrastructure, including reliable electricity, internet connectivity, cloud computing resources, and sensor networks. These prerequisites are not uniformly available, particularly in rural and low-income regions (Vinuesa et al., 2020).

For example, real-time flood early warning systems depend on dense networks of sensors and robust communication links, which may be lacking in high-risk areas. Without adequate infrastructure, even the most advanced AI models cannot deliver timely and actionable insights.

Edge computing and decentralized AI architectures offer promising solutions to some of these infrastructure challenges by enabling data processing closer to the source. However, these technologies are still emerging and often involve trade-offs in computational power and energy consumption.

Ethical Risks and Unintended Consequences

AI systems can inadvertently introduce new risks or exacerbate existing vulnerabilities. For instance, predictive policing algorithms based on environmental factors like heatwaves or urban decay may disproportionately target marginalized communities, leading to unjust outcomes (Barocas et al., 2019).

Similarly, poorly designed evacuation algorithms that prioritize efficiency over equity might systematically disadvantage elderly, disabled, or economically disadvantaged populations. In the context of environmental hazard mitigation, where rapid, large-scale decisions are often necessary, these ethical pitfalls can have profound real-world impacts.

Unintended consequences also arise from model feedback loops. For example, if resource allocation decisions based on AI predictions cause certain areas to be neglected systematically, long-term vulnerabilities may be entrenched rather than alleviated.

Dependence on AI and Loss of Traditional Knowledge

An overreliance on AI systems risks displacing traditional ecological knowledge and community-based disaster preparedness strategies. Indigenous and local knowledge systems have historically played crucial roles in managing environmental risks (Hiwasaki et al., 2014).

If communities are encouraged to depend solely on AI-based warnings and interventions, valuable traditional practices — such as early animal migration observations or local flood markers — may be lost. This not only diminishes cultural heritage but can also reduce the redundancy and resilience of disaster management systems.

Integrating AI with local knowledge, rather than replacing it, is therefore essential for sustainable and culturally sensitive hazard mitigation efforts.

Cost, Resource Requirements, and Sustainability

Developing, deploying, and maintaining AI systems for environmental resilience can be costly. Training sophisticated models demands significant computational resources, which translates to high energy consumption and carbon footprints — a paradox when the goal is often to protect ecosystems and combat climate change (Strubell et al., 2019).

Moreover, continuous updating and retraining of models are necessary as environmental conditions and risk factors evolve. These ongoing costs can be prohibitive for resource-constrained governments and organizations, potentially widening the resilience gap between wealthy and poorer regions.

Advancements in green AI — focusing on reducing the environmental impact of machine learning — are critical to addressing this sustainability challenge (Schwartz et al., 2020).

Regulatory and Policy Gaps

The regulatory environment for AI applications in environmental hazard mitigation is underdeveloped. Few countries have established clear guidelines for the ethical and legal deployment of AI in disaster management contexts. This creates uncertainty for developers and operators and raises risks for affected populations (Jobin et al., 2019).

Without robust regulatory frameworks, AI systems may be deployed without adequate testing, validation, or oversight. This is especially concerning given the high-stakes nature of environmental hazard response, where errors can have catastrophic consequences. Developing coherent, adaptive, and enforceable AI governance policies is therefore a pressing need to ensure the safe and effective use of AI in this domain.

POLICY RECOMMENDATIONS

As artificial intelligence (AI) rapidly integrates into environmental hazard mitigation, policymakers, scientists, and community leaders must act collectively to shape its deployment for the greatest public benefit. Strategic policies can amplify AI's strengths, address its limitations, and ensure that the future of environmental resilience is inclusive, ethical, and sustainable. This section outlines key policy recommendations and offers a forward-looking perspective on how AI can help build a safer, more resilient world.

Building Inclusive and Equitable AI Systems

AI applications in hazard mitigation must prioritize equity and inclusion from the outset. Historically marginalized and vulnerable communities often bear the brunt of environmental hazards (Islam & Winkel, 2017). To prevent the reproduction of these disparities, policies must require inclusive data practices, participatory design approaches, and equity-focused outcome evaluations. Stakeholder engagement frameworks should mandate the meaningful involvement of indigenous groups, low-income communities, and marginalized populations in AI system design and deployment (Vinuesa et al., 2020). Funding mechanisms must also prioritize initiatives that explicitly target resilience-building for at-risk populations. For example, participatory mapping and community-based data collection can ensure that AI models reflect local realities and needs, rather than imposing one-size-fits-all solutions that may be ineffective or harmful (Peters et al., 2017).

Promoting Data Sovereignty and Open Data Standards

Given AI's dependence on large, high-quality datasets, clear policies on data sovereignty, ownership, and sharing are essential. Local communities and governments should retain control over their environmental and hazard-related data while benefiting from open-data collaborations that enhance modeling capabilities. International frameworks, similar to the FAIR Data Principles (Findable, Accessible, Interoperable, Reusable), should be adopted to standardize data-sharing practices (Wilkinson et al., 2016). Moreover, policies must ensure that data is collected ethically, respecting privacy rights and community consent. Creating global open-access hazard data platforms — funded and governed by international bodies like the UN or World Bank — could democratize access to critical information, especially for resource-constrained countries.

Fostering Transparency, Explainability, and Accountability

Policy must enshrine transparency, explainability, and accountability as core principles in AI governance. This is particularly important in environmental hazard contexts where decisions affect human safety and well-being. Mandatory model documentation, audit trails, and interpretability tools should be required for AI systems deployed in disaster management (Morley et al., 2021). These mechanisms can help ensure that both experts and laypersons understand how risk assessments, predictions, or response recommendations are generated. Additionally, clear lines of accountability must be established. If an AI-based system fails or causes harm, regulatory frameworks must identify who is responsible: the developer, the operator, the deploying agency, or another entity.

Advancing Research in Green AI and Sustainable Technologies

Given the high energy consumption of AI training and inference, governments must incentivize the development of "green AI" — environmentally sustainable AI methods that minimize computational and carbon footprints (Schwartz et al., 2020). Funding agencies should prioritize research on low-energy machine learning models, edge computing solutions for remote hazard monitoring, and energy-efficient hardware. Tax credits or grants could encourage private sector investment in these technologies. Further, environmental impact assessments should be made mandatory for large-scale AI deployments in hazard mitigation, ensuring that their operation does not inadvertently contribute to the environmental problems they aim to solve.

Strengthening Infrastructure for AI Deployment

AI's effectiveness hinges on reliable supporting infrastructure: sensors, communication networks, cloud storage, and computational resources. Policymakers must invest heavily in building resilient digital and physical infrastructure, particularly in vulnerable and underserved areas (Mulligan et al., 2020). Public-private partnerships can play a vital role in scaling sensor networks for flood monitoring, wildfire detection, and early warning systems. Policies should encourage shared infrastructure models, where governments, NGOs, and the private sector collaborate on maintaining essential networks. In remote regions, decentralized, off-grid AI systems powered by renewable energy (such as solar-powered sensors) offer promising pathways to reduce dependency on centralized infrastructure.

Establishing Ethical Standards and Certification Frameworks

Global ethical standards for AI in hazard mitigation must be developed, adapted to local contexts, and enforced through certification systems. These standards should cover fairness, bias mitigation, privacy protection, and human oversight. Certification bodies — potentially modeled on existing standards organizations like ISO or IEEE — could assess AI systems before deployment, ensuring that they meet rigorous ethical and technical benchmarks (Jobin et al., 2019). Moreover, continuous monitoring mechanisms are crucial to detect drift in model behavior over time and to prevent ethical lapses as environmental conditions and datasets evolve.

Preparing the Workforce and Upskilling Communities

Policymakers must invest in human capital to ensure that the benefits of AI in environmental resilience are widely distributed. This includes training a new generation of AI-literate emergency managers, environmental scientists, and public health officials. Community education programs are equally important. Residents of high-risk areas should be trained to understand, interact with, and critique AI-based hazard warning systems. Doing so enhances community ownership of resilience efforts and reduces technology-related distrust (Pasquier et al., 2020). Universities and vocational training institutions should embed interdisciplinary curricula combining AI, environmental science, ethics, and disaster management, preparing future leaders for the complex challenges ahead.

Enhancing International Collaboration

Environmental hazards transcend national borders; so must the solutions. Stronger international cooperation is essential for knowledge sharing, standard setting, and coordinated response to transboundary disasters like wildfires, pandemics, and climate-induced migration. Multilateral initiatives, such as the Sendai Framework for Disaster Risk Reduction (UNDRR, 2015), should explicitly integrate AI strategies into their action plans. Global research consortia can facilitate cross-country learning on effective AI practices, particularly for rare or emerging hazards. Moreover, international aid programs should increasingly include technological capacity-building components, helping lower-income nations leverage AI in their own hazard mitigation efforts.

Future Outlook: Towards Adaptive and Resilient AI Ecosystems

Looking ahead, the future of AI in environmental hazard mitigation will likely evolve toward more adaptive, decentralized, and participatory ecosystems. Advances in federated learning could allow models to be trained across multiple sites without centralized data pooling, preserving privacy while enhancing diversity (Yang et al., 2019). The integration of AI with emerging technologies such as the Internet of Things (IoT), blockchain, and satellite remote sensing will enable real-time, hyper-localized hazard detection and response systems. However, realizing this future requires sustained commitment to ethical principles, inclusivity, and environmental stewardship. Rather than replacing human judgment, AI must augment human capacity to foresee, prepare for, and recover from environmental crises. If harnessed wisely, AI offers not just a technological toolkit, but a transformative force to reimagine resilience for a safer and more sustainable world.

CONCLUSION

The convergence of AI and environmental hazard mitigation represents one of the most significant opportunities of our time to enhance societal resilience and safeguard ecosystems. As this chapter has shown, AI technologies are already playing a transformative role across the hazard management lifecycle — from prediction and early warning to response coordination and long-term recovery. Yet, as with all powerful tools, the realization of AI's full potential demands intentional design, ethical governance, inclusive engagement, and ongoing vigilance.

Environmental hazards are intensifying in frequency and severity, propelled by factors such as climate change, rapid urbanization, and industrial expansion (IPCC, 2022). Traditional approaches to disaster management, while critical, are increasingly insufficient to address the complexity and scale of contemporary risks. AI offers capabilities that can dramatically augment human decision-making: it can synthesize vast and diverse data streams, identify subtle patterns and anomalies, generate predictive models, and autonomously trigger responsive actions.

For example, machine learning algorithms can predict flood risks with unprecedented precision by integrating meteorological, topographical, and hydrological data (Mosavi et al., 2018). Remote sensing combined with AI-driven image analysis enables real-time wildfire detection and monitoring (Kobayashi et al., 2019). In the aftermath of disasters, AI-powered tools can rapidly assess damages, optimize logistics for relief distribution, and support psychological health interventions through smart applications (Hong et al., 2021).

However, the integration of AI into environmental hazard mitigation is not without significant challenges. Technical issues such as data scarcity, model generalizability, and system robustness under novel conditions can impair the reliability of AI-driven insights. Moreover, ethical concerns around data privacy, algorithmic bias, accountability, and equitable access must be confronted head-on to prevent exacerbating existing vulnerabilities and inequalities.

REFERENCES

Alexander, D. E. (2020). *Natural disasters*. Routledge., DOI: 10.4324/9781003071372

Apte, J. S., Messier, K. P., Gani, S., Brauer, M., Kirchstetter, T. W., Lunden, M. M., & Marshall, J. D. (2017). High-resolution air pollution mapping with Google Street View cars: Exploiting big data. *Environmental Science & Technology*, *51*(12), 6999–7008. DOI: 10.1021/acs.est.7b00891 PMID: 28578585

Baker, J., Ebrahim, Z., & Zarina, O. (2019). *Data challenges for climate resilience: Empowering action through better information*. World Bank Group.

Barocas, S., Hardt, M., & Narayanan, A. (2019). *Fairness and machine learning: Limitations and opportunities*.https://fairmlbook.org

Batty, M. (2018). Digital twins. *Environment and Planning. B, Urban Analytics and City Science*, *45*(5), 817–820. DOI: 10.1177/2399808318796416

Beck, M. B., Balbi, S., & Lanzanova, D. (2020). Bridging the gap between data and decision-making: Leveraging artificial intelligence in support of sustainable development. *Environmental Modelling & Software*, *132*, 104808. DOI: 10.1016/j.envsoft.2020.104808

Boudreau, M. E., DeVries, B., & Wessels, K. (2021). Transfer learning for environmental hazard mapping: Challenges and opportunities. *Remote Sensing*, *13*(12), 2345. DOI: 10.3390/rs13122345

Burrell, J. (2016). How the machine 'thinks': Understanding opacity in machine learning algorithms. *Big Data & Society*, *3*(1), 2053951715622512. DOI: 10.1177/2053951715622512

Calo, R. (2016). Robotics and the lessons of cyberlaw. *California Law Review*, *103*(3), 513–563. DOI: 10.15779/Z38XX96

Cheng, X., Hu, Y., & Yu, G. (2022). AI in wildfire prediction and monitoring: A review. *International Journal of Wildland Fire*, *31*(7), 617–627. DOI: 10.1071/WF21035

Chollet, F. (2021). *Deep learning with Python* (2nd ed.). Manning Publications.

Cutter, S. L. (2006). The geography of social vulnerability: Race, class, and catastrophe. *Social Science Quarterly*, *87*(2), 286–291. DOI: 10.1111/j.1540-6237.2006.00384.x

Cutter, S. L., Burton, C. G., & Emrich, C. T. (2008). Disaster resilience indicators for benchmarking baseline conditions. *Journal of Homeland Security and Emergency Management*, *5*(1). Advance online publication. DOI: 10.2202/1547-7355.1732

Cutter, S. L., Burton, C. G., & Emrich, C. T. (2010). Disaster resilience indicators for benchmarking baseline conditions. *Journal of Homeland Security and Emergency Management*, *7*(1). Advance online publication. DOI: 10.2202/1547-7355.1732

Elish, M. C. (2019). Moral crumple zones: Cautionary tales in human-robot interaction. *Engaging Science, Technology, and Society*, *5*, 40–60. DOI: 10.17351/ests2019.260

Fjeld, J., Achten, N., Hilligoss, H., Nagy, A., & Srikumar, M. (2020). Principled artificial intelligence: Mapping consensus in ethical and rights-based approaches to principles for AI. *Berkman Klein Center Research Publication No. 2020-1*. DOI: 10.2139/ssrn.3518482

Floridi, L., Cowls, J., Beltrametti, M., Chatila, R., Chazerand, P., Dignum, V., Luetge, C., Madelin, R., Pagallo, U., Rossi, F., Schafer, B., Valcke, P., & Vayena, E. (2018). AI4People—An ethical framework for a good AI society: Opportunities, risks, principles, and recommendations. *Minds and Machines*, *28*(4), 689–707. DOI: 10.1007/s11023-018-9482-5 PMID: 30930541

Folke, C., Carpenter, S. R., Walker, B., Scheffer, M., Chapin, T., & Rockström, J. (2010). Resilience thinking: Integrating resilience, adaptability and transformability. *Ecology and Society*, *15*(4), 20. DOI: 10.5751/ES-03610-150420

Gao, L., Zhang, L., & Chen, X. (2022). Deep learning in disaster emergency response: Applications and challenges. *Natural Hazards*, *113*(2), 1375–1400. DOI: 10.1007/s11069-022-05124-w

Goodfellow, I., Bengio, Y., & Courville, A. (2016). *Deep learning*. MIT Press.

Hiwasaki, L., Luna, E., Syamsidik, , & Shaw, R. (2014). Process for integrating local and indigenous knowledge with science for hydro-meteorological disaster risk reduction and climate change adaptation in coastal and small island communities. *International Journal of Disaster Risk Reduction*, *10*, 15–27. DOI: 10.1016/j.ijdrr.2014.07.007

Hong, Y., Liu, Q., & Hu, X. (2021). AI technologies for disaster management. *International Journal of Disaster Risk Reduction*, *64*, 102495. DOI: 10.1016/j.ijdrr.2021.102495

Intergovernmental Panel on Climate Change (IPCC). (2023). *Climate Change 2023: Synthesis Report*. Cambridge University Press., DOI: 10.1017/9781009325844

IPCC. (2022). Climate Change 2022: Impacts, Adaptation and Vulnerability. Contribution of Working Group II to the Sixth Assessment Report of the Intergovernmental Panel on Climate Change. Cambridge University Press.

Islam, S. N., & Winkel, J. (2017). Climate change and social inequality. *DESA Working Paper No. 152*. https://doi.org/DOI: 10.18356/9dbcb5b4-en

Jobin, A., Ienca, M., & Vayena, E. (2019). The global landscape of AI ethics guidelines. *Nature Machine Intelligence*, *1*(9), 389–399. DOI: 10.1038/s42256-019-0088-2
Jobin, A., Ienca, M., & Vayena, E. (2019). The global landscape of AI ethics guidelines. *Nature Machine Intelligence*, *1*(9), 389–399. DOI: 10.1038/s42256-019-0088-2
Kobayashi, T., Tanaka, K., & Kobayashi, K. (2019). Wildfire detection from satellite images using deep convolutional neural networks. *Remote Sensing*, *11*(23), 2874. DOI: 10.3390/rs11232874

Kreibich, H., Bubeck, P., Van Vliet, M., & De Moel, H. (2017). A review of damage-reducing measures to manage fluvial flood risks in a changing climate. *Mitigation and Adaptation Strategies for Global Change*, *22*(5), 963–989. DOI: 10.1007/s11027-016-9727-5

Liu, Y., Liu, X., & Li, Y. (2022). Advances in AI-powered disaster risk reduction: A review. *International Journal of Disaster Risk Science*, *13*(1), 1–18. DOI: 10.1007/s13753-022-00379-x

Maas, P., Iyer, S., Gros, A., Park, W., & McGorman, L. (2020). Facebook Disaster Maps: Aggregate insights for crisis response & recovery. *Proceedings of the 27th ACM SIGSPATIAL International Conference on Advances in Geographic Information Systems*, 579–582. https://doi.org/DOI: 10.1145/3347146.3359332

Matin, N., Forrester, J., & Ensor, J. (2018). What is equitable resilience? *World Development*, *109*, 197–205. DOI: 10.1016/j.worlddev.2018.04.020 PMID: 30177865

Mehrabi, N., Morstatter, F., Saxena, N., Lerman, K., & Galstyan, A. (2021). A survey on bias and fairness in machine learning. *ACM Computing Surveys*, *54*(6), 1–35. DOI: 10.1145/3457607

Mori, J., Ohta, Y., & Saito, T. (2019). Tsunami early warning and real-time monitoring using big data and AI technologies. *Pure and Applied Geophysics*, *176*(5), 2021–2036. DOI: 10.1007/s00024-018-2010-1

Morley, J., Floridi, L., Kinsey, L., & Elhalal, A. (2021). From what to how: An initial review of publicly available AI ethics tools, methods and research to translate principles into practices. *Science and Engineering Ethics*, *27*(1), 1–19. DOI: 10.1007/s11948-021-00282-w PMID: 33492516

Mosavi, A., Ozturk, P., & Chau, K. W. (2018). Flood prediction using machine learning models: Literature review. *Water (Basel)*, *10*(11), 1536. DOI: 10.3390/w10111536

Mulligan, C., Olivadoti, L., & Buck, C. (2020). Edge AI: The convergence of AI and edge computing. *Journal of Information Technology Teaching Cases*, *10*(1), 31–36. DOI: 10.1177/2043886920902736

Narayanan, A., & Shmatikov, V. (2008). Robust de-anonymization of large sparse datasets. *IEEE Symposium on Security and Privacy*, 111–125. DOI: 10.1109/SP.2008.33

O'Neil, C. (2016). *Weapons of math destruction: How big data increases inequality and threatens democracy*. Crown Publishing Group.

OECD. (2019). *OECD Principles on Artificial Intelligence*. Retrieved from https://www.oecd.org/going-digital/ai/principles/

Pasquier, T., Lau, M. K., Trippas, J. R., & Jay, C. (2020). Artificial intelligence for disaster response and resilience: Recent advances and future directions. *IEEE Security and Privacy*, *18*(4), 29–36. DOI: 10.1109/MSEC.2020.2998726

Patel, H., Jain, V., & Sharma, A. (2019). Operation Blue Sky: An AI-powered flood forecasting system for India. *IBM Journal of Research and Development*, *63*(6), 1–12. DOI: 10.1147/JRD.2019.2952187

Perrow, C. (2011). *Normal accidents: Living with high-risk technologies* (2nd ed.). Princeton University Press. DOI: 10.2307/j.ctt7srgf

Peters, L., Thaler, T., Hanger-Kopp, S., & Seebauer, S. (2017). Understanding the dynamics of risk perception and community resilience. *Natural Hazards*, *86*(1), 353–376. DOI: 10.1007/s11069-016-2705-3

Rolnick, D., Donti, P. L., Kaack, L. H., Kochanski, K., Lacoste, A., Sankaran, K., & Bengio, Y. (2022). Tackling climate change with machine learning. *Communications of the ACM*, *65*(10), 47–59. DOI: 10.1145/3485128

Rolnick, D., Donti, P. L., Kaack, L. H., Kochanski, K., Lacoste, A., Sankaran, K., & Bengio, Y. (2022). Tackling climate change with machine learning. *Communications of the ACM*, *65*(10), 47–59. DOI: 10.1145/3485128

Rudin, C. (2019). Stop explaining black box machine learning models for high stakes decisions and use interpretable models instead. *Nature Machine Intelligence, 1*(5), 206–215. DOI: 10.1038/s42256-019-0048-x PMID: 35603010

Russell, S., & Norvig, P. (2021). *Artificial Intelligence: A Modern Approach* (4th ed.). Pearson.

Samek, W., Wiegand, T., & Müller, K.-R. (2017). Explainable artificial intelligence: Understanding, visualizing and interpreting deep learning models. *arXiv preprint* arXiv:1708.08296.

Schwartz, R., Dodge, J., Smith, N. A., & Etzioni, O. (2020). Green AI. *Communications of the ACM, 63*(12), 54–63. DOI: 10.1145/3381831

Smith, K. (2013). *Environmental hazards: Assessing risk and reducing disaster* (6th ed.). Routledge., DOI: 10.4324/9780203805305

Strubell, E., Ganesh, A., & McCallum, A. (2019). Energy and policy considerations for deep learning in NLP. *Proceedings of the 57th Annual Meeting of the Association for Computational Linguistics*, 3645–3650. DOI: 10.18653/v1/P19-1355

Subramanian, V., Holzer, A., & Weis, T. M. (2021). Building resilient AI: Principles and challenges. *AI and Ethics, 1*(1), 43–53. DOI: 10.1007/s43681-020-00003-1

UNDRR. (2015). Sendai Framework for Disaster Risk Reduction 2015–2030. United Nations Office for Disaster Risk Reduction. https://www.undrr.org/publication/sendai-framework-disaster-risk-reduction-2015-2030

UNESCO. (2021). *Recommendation on the Ethics of Artificial Intelligence.* Retrieved from https://unesdoc.unesco.org/ark:/48223/pf0000381137

United Nations Office for Disaster Risk Reduction (UNDRR). (2015). *Sendai Framework for Disaster Risk Reduction 2015–2030.* https://www.undrr.org/publication/sendai-framework-disaster-risk-reduction-2015-2030

United Nations Office for Disaster Risk Reduction (UNDRR). (2019). *Global Assessment Report on Disaster Risk Reduction.* https://gar.undrr.org

Vinuesa, R., Azizpour, H., Leite, I., Balaam, M., Dignum, V., Domisch, S., & Nerini, F. F. (2020). The role of artificial intelligence in achieving the Sustainable Development Goals. *Nature Communications, 11*(1), 233. DOI: 10.1038/s41467-019-14108-y PMID: 31932590

Wilkinson, M. D., Dumontier, M., Aalbersberg, I. J., Appleton, G., Axton, M., Baak, A., Blomberg, N., Boiten, J.-W., da Silva Santos, L. B., Bourne, P. E., Bouwman, J., Brookes, A. J., Clark, T., Crosas, M., Dillo, I., Dumon, O., Edmunds, S., Evelo, C. T., Finkers, R., & Mons, B. (2016). The FAIR Guiding Principles for scientific data management and stewardship. *Scientific Data*, *3*(1), 160018. DOI: 10.1038/sdata.2016.18 PMID: 26978244

Winfield, A. F., & Jirotka, M. (2018). Ethical governance is essential to building trust in robotics and artificial intelligence systems. *Philosophical Transactions. Series A, Mathematical, Physical, and Engineering Sciences*, *376*(2133), 20180085. DOI: 10.1098/rsta.2018.0085 PMID: 30323000

Yang, Q., Liu, Y., Chen, T., & Tong, Y. (2019). Federated machine learning: Concept and applications. *ACM Transactions on Intelligent Systems and Technology*, *10*(2), 12. DOI: 10.1145/3298981

Chapter 7
Mapping Urban Planning and Sustainability: A Bibliometric Perspective

Neeru Sidana
https://orcid.org/0000-0003-1087-4214
Amity University, Noida, India

Ajay Sidana
Amity University, Noida, India

Dewanshi Ghosh
https://orcid.org/0009-0008-6610-9208
Amity University, Noida, India

Rohit Sood
https://orcid.org/0000-0002-4127-7781
Lovely Professional University, India

ABSTRACT

The purpose of studying the nexus between urban planning and sustainability through a bibliometric perspective is to systematically analyze and understand the existing body of scholarly literature on this topic. This paper aims to review articles related to urban planning and sustainability with the help of bibliometric analysis. A total number of 899 documents were drawn from the Scopus database out of which 377 articles were selected. To accomplish the overview of Urban planning and sustainability Research, bibliometric techniques such as performance analysis and scientific mapping were applied. The data was also organized, analyzed, and presented using the Bibliometrix package. The study found that research on Urban planning and sustainability Research has a significant growth rate of -15.39 percent. It was

DOI: 10.4018/979-8-3373-3206-2.ch007

also discovered that DESLATTE A has produced the most influential articles in this field and Geojournal was the top-ranked journal. The findings also highlighted that USA, India and China had the highest rates of research productivity in this field.

INTRODUCTION

The increasing rate of urbanization in the twenty-first century has put great pressure on cities to strike a balance between expansion and sustainability. More than half of the world's population already lives in cities, and estimates indicate that this tendency will only continue. As a result, urban planning has expanded beyond physical infrastructure to include environmental stewardship, social equality, and economic resilience (Hamman, 2017) & (Sharifi, 2021). Understanding how research in this area has evolved is critical, and bibliometric analysis offers a valuable tool for mapping these intellectual paths, identifying patterns, and guiding future study.

The convergence of sustainability and urban planning has resulted in an extensive and multidisciplinary literature base. Early studies focused on energy usage and efficiency in urban environments. (Cicea, 2020) Did a bibliometric assessment to demonstrate how energy-efficient planning and renewable energy integration become important to urban strategy. Similarly, (Cui, 2018) investigated the idea of urban metabolism, providing insights into how cities operate as dynamic systems of resource exchange.

Another issue of vital importance for sustainable urban systems is transport. (Ince, 2025) analysed global research trends related to sustainable mobility, with emphasis on low-emission transportation as well as walking-oriented city landscapes. GIS is also changing the approach to planning. (Celik & Sekeroglu, 2023) emphasize the Significance of GIS to visualize and optimize land use and demonstrates a shift towards a data driven decision making.

The concept of smart and sustainable cities has been widely accepted. (Janik, Kiesl, Twoich, Sosinska, Wu & Netter, 2020) showed the confluence of digital innovation and environmental ambitions and how governance, ICT, and sustainability were embedded in urban frames. (Zaidi et al., 2023) studied how AI could contribute to the design of Smart City systems by reducing energy consumption, controlling traffic or delivering urban services.

Urban resilience has become more popular as cities grapple with more and more climate-related hazards. (Guo et al., 2021) focused on the robustness of infrastructure, disaster mitigation, and adaptive urban design, whereas (Yang et al., 2023) provided a territorial approach to resilient planning that considered the integration of spatial governance and long-term sustainability.

Land use is an important aspect of planning debates. (Lv et al., 2021) conducted a bibliometric analysis which brought attention on the issue of urban growth, and on how crucial integrated land-use policy is to address it. (Sheng et al., 2022) confirmed this through studying institutional arrangements of urban land management.

The social factors of urban sustainable have also been researched. (Liang et al., 2022) and (Razia & Ah, 2022) explored the dimensions of inclusion and fairness by emphasizing the need for equitable access to resources, participation and cultural cohesiveness for the sustainable future of urban systems.

Environmental infrastructure, particularly green infrastructure like parks, roofs and walls, is popular. (Bonakdar et al., 2024) have also examined the benefits of green infrastructures in terms of air quality and the enhancement of climate resilience. Green Roofs: (Li et al., 2024) studied the green-roof turning point, while (Wang et al., 2019) depicted the harmonious interplay between BIM and GIS for environmentally friendly construction.

Diversity of academic disciplines among studies in the urban health space is on the rise. (Jia et al., 2021) investigated shifts in conceptions of healthy urban development over a forty-year period, while (Chen et al., 2021) related urban form to thermal comfort. The relationship between urban settings and public health and well-being was highlighted by (Van Winkle et al., 2022).

Increasingly participative planning models are being designed that are responding to citizens interaction possibilities through digital tools. As reported by (Liao et al., 2019), crowdsourcing is such a frequent practice in planning (showing how participative power contributes to dynamic urban governance and design improvements).

Tourism and sustainable urban development have also met. The transformation of urban tourism has been investigated by (Hocevar & Bartol, 2021) and (Krabokoukis & Polyzos, 2023) analysis on heritage conservation and sustainability. Asif et al. (2025) discussed the increasing importance of the metaverse in sustainable tourism and provided creative thoughts for low-impact urban experiences.

Of late, bibliometric methods have contributed to assessing the overall liveability and the quality of metropolitan areas. Molaei Qelichi et al. (2025) traced the development of urban liveability research to integrate concepts of safety, access, and further quality of life components, while D'Alpaos and Andreolli (2020) used multicriteria evaluations to assess city performance.

Moving forward, a holistic understanding can be synthesized through integrating long-term historical and conceptual analyses. In a recent study, Qian and colleagues (2023) conducted a thorough review of six decades worth of literature and highlighted the interdisciplinary foundations of sustainable cities. They integrated work tackling the development of sustainability in urban planning over 35 years (Fu & Zhang, 2017). Going beyond the urban core, the often neglected planning needs of the peri-urban and rural-urban fringe were analysed by Sareen and Haque in their 2023 study.

Where Oliva et al. (2022) connected city branding to the outcomes of urban development, Wang et al. (2022) and Marvuglia et al. (2020) deepened the understanding regarding the sustainability transitions and assessment methods. Wagner (2019) focused on the port cities and the specific metropolitan sustainability concerns.

In conclusion, there is a wealth of diverse research on urban planning and sustainability that covers topics such as energy, transportation, land use, health, governance, and new technology. By mapping research clusters, identifying significant authors and publications, and spotting new trends, bibliometric analysis aids in the mapping of this complexity. This analytical approach offers a strategic tool and a retrospective viewpoint for directing sustainable urban futures as urban issues increase.

LITERATURE REVIEW

Evolution of Urban Planning and Sustainability Research

Rigid zoning procedures have given way to an integrated form of urban planning that takes sustainability's social, economic, and environmental aspects into account. While new research shows a multifaceted view of how cities run sustainably, earlier studies focused on urban design and infrastructure (Fu & Zhang, 2017) (Wang, Ho & Fu, 2019), (Sharifi, 2021). Researchers have monitored the transition from physical planning to inclusive, participatory governance models using bibliometric methodologies (Hamman, 2017), (Liang et al., 2022). Important turning points include the increasing focus on energy efficiency in urban settings (Cicea, 2020) and resilience (Guo et al., 2021), which indicate the relationship between sustainability and ecological and technical innovation.

Scholars may now observe trends, keyword co-occurrences, and institutional partnerships in sustainability literature thanks to bibliometric analysis, especially with tools like CiteSpace and VOSviewer (Chen, Wang & Zhou, 2021), (Li et al., 2024). This technologies have been used to map the spatial and temporal aspects of research, revealing popular subjects and new areas such as green infrastructure and smart cities (Janik, Ryszko & Szafraniec, 2020), (Sokolova et al., 2024). Research like that conducted by (Wang et al.,2022) demonstrates how bibliometric techniques offer a critical perspective for comprehending paradigm shifts in the discourse around sustainability.

Smart and Sustainable Urban Systems

Under the general heading of smart cities, there is a lot of discussion on the relationship between sustainability and digital technology. The application of AI, IoT, and data-driven technologies in improving urban systems is currently the subject of a sizable body of work (Zaidi et al., 2023), (Janik et al., 2020). Bibliometric mapping has shown that e-governance, energy management, and intelligent transportation are among the most popular themes. According to (İnce, 2025) and (Soliman et al., 2021), smart infrastructure changes urban government paradigms by promoting both operational efficiency and environmental aims.

In urban settings, environmental sustainability entails consideration of green areas, land use trends, and urban metabolism. (Cui, 2018) provides information on how cities might function as metabolic systems that balance inflows and outflows—a notion that has been thoroughly examined via the use of bibliometric analysis. Research on green roofs and green infrastructure, which greatly enhance ecosystem services in crowded cities, is the topic of studies by (Sokolova et al., 2024) and (Li et al., 2024). Furthermore, land use continues to be a persistent challenge, with sprawl and fragmentation being cited as barriers to sustainability (Sheng et al., 2022), (Siyavuş & Aydın, 2022).

Urban Resilience and Risk Management

The idea of resilience has developed to address the ability of urban systems to adjust to shocks and pressures, especially when faced with pandemics, sociopolitical strife, and climate change. According to (Guo et al., 2021) and (Yang et al., 2023), resilience is currently a major topic in the research on urban planning. According to bibliometric evaluations, urban resilience is becoming more and more popular, with an emphasis on risk-sensitive land use, infrastructural resilience, and adaptive governance (Pascariu et al., 2023).

The concept of inclusive urban planning, which prioritizes social sustainability, fairness, and accessibility, is becoming more popular. In their methodical analysis of these factors, (Liang et al., 2022) pointed out that inclusive cities are those that give priority to underserved populations, protect their right to movement, and encourage public involvement. Similarly, (Molaei Qelichi et al., 2025) used bibliometric techniques to track down studies on livability, uncovering themes linked to social cohesiveness, walkability, and housing quality. Together, these factors influence long-term sustainability and urban well-being.

Role of Governance, Participation, and Policy Integration

Multilevel governance, participatory planning, and crowdsourcing are changing the way municipal policies are created and carried out. (Liao et al., 2019) map how public participation and digital platforms might improve urban decision-making. According to (Wang at al., 2019), governance structures today prioritize collaborative models that incorporate local, regional, and international policy. Stakeholder engagement is becoming increasingly important, which is in line with sustainability objectives and democratic planning procedures.

Tourism and transportation have become industries that are closely related to sustainable urban design. (Krabokoukis & Polyzos, 2023) and (Asif et al., 2025) demonstrate the effects of tourism on infrastructure, cultural landscapes, and land use. Through a worldwide trend study, (İnce, 2025) highlighted innovations and difficulties in transportation, particularly in the context of sustainable mobility. With a discernible progression in the study of renewable integration in urban systems, energy continues to be a fundamental pillar of sustainability research (Cicea, 2020), (Marvuglia et al., 2020).

Spatial Planning and Peri-Urban Challenges

Peri-urban communities face difficulties as a result of urban expansion that frequently transcends conventional limits. A thorough review of planning concerns at the urban-rural interface is given by (Sareen & Haque, 2023). Their research points out a discrepancy in how land use governance and policy are coordinated, which has an impact on sustainability transitions. According to (Qian at al.,2023), spatial planning studies concentrate on incorporating socioeconomic objectives and environmental protections into zoning systems.

With more studies being conducted on the effects of urban design on both physical and mental health, health is becoming a more prominent lens in sustainability studies. According to Van (Winkle et al., 2022), bibliometric clusters on urban welfare currently focus heavily on issues including noise, pollution, and urban green space. In a similar (vein et al., 2021) examined forty years of research on healthy cities and saw a change in emphasis from illness prevention to lifestyle-focused urban planning. (D'Alpaos & Andreolli, 2020) examined the concept of urban quality, which includes both subjective perceptions and objective infrastructural quality.

Data Analysis

- **Annual Scientific Production**

An effective way to evaluate the foundational, developing, and mature stages of a research topic is to look at the pattern of yearly publishing activity. The annual distribution of research papers from 2016 to 2025 that highlight the relationship between sustainability and urban planning in the framework of lifelong learning is depicted in Figure 1. Elsevier B.V.'s Scopus scientific database served as the source of the data used in this research. The graph shows that scientific production started off at about 27 articles in 2016 and then significantly increased to about 46 articles in 2017. A significant drop to about 29 articles in 2018 followed this.

After a modest rebound in 2019, the number of publications increased steadily, reaching its maximum point in 2022 with about 59 pieces. After 2022, there was a drop in 2023 with around 39 articles, and then a slight increase in 2024 with about 51 articles. These graphs are useful analytical tools for assessing the yearly production of scientific publications in many fields of study. The yearly growth rate of publications in the topic area under investigation is determined to be -15.39% based on this trend analysis, which shows a decrease in total research output throughout the given time frame.

Figure 1. Annual scientific production from 2016- 25.

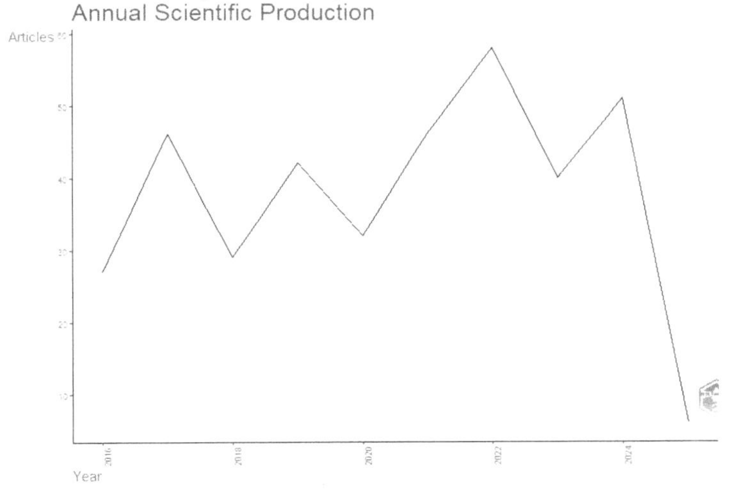

- **Citations Per Year**

Citation analysis is one method that may be used to assess the quality and significance of a scientific work. Figure 2 displays the citation trends from 2016 to 2025. These images show the differences in citations that occur within that area.

This graphic's data may be useful for assessing the influence and impact of published scientific publications in the field. This image may also be used to identify noteworthy years or occasions that affect the number of citations for the domain.

According to Figure 2, which displays a sharp increase in citations between 2017 and 2020 and a gradual fall starting in 2020, the average number of citations per document was 25.45. Given the availability of public data, this increase may be attributed to notable advancements and recognition in the fields of urban planning and resource management's sustainable culture during the previous several years.

Figure 2. Citations per year

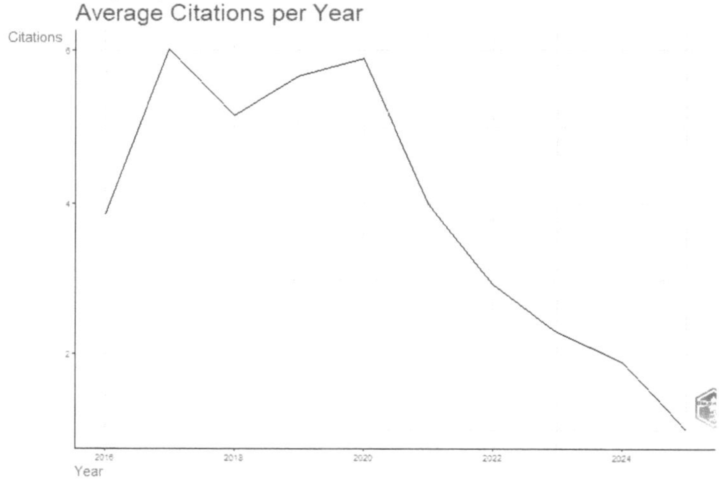

- **Most Relevant Sources**

Regarding Figure 3, the total number of documents for each source is shown in this line graph. Regarding the overall quantity of articles with sources, each element in the picture has a distinct function. These sources include scholarly publications on sustainability, human resource management, leadership, and technology. There are ten labels on the Y axis.

The number of publications received from Eastern sources is shown by the data displayed by the blue dart. The number of documents and the source name are represented by the X and Y coordinates of each point. There is a dedicated department for research and development on urbanization and sustainability at the University of California, Arizona State University, Beijing Normal University, and Minnesota.

The sources that are most pertinent are "GEOJOURNAL", "ENVIRONMENTAL, DEVELOPMENT and SUSTAINABILITY", "ENVIRONMENTAL SCIENCE AND POLLUTION RESEARCH", "INTERNATIONAL JOURNAL OF URBAN SUSTAINABLE DEVELOP", "LANDSCAPE AND URBAN PLANNING", "LOCAL ENVIRONMENT", "PROCEEDING TO THE INSTITUTE OF CIVIL ENGINEERS", "ENVIRONEMT AND PLANNING B; URBAN ANALYTICS AND CI", "ENVIRONMENT AND URBANIZATION ASIA", & "INTERNATIONAL JOURNAL OF SUSTAINABILITY DEVELOPMENT A", the sources listed in Figure 3. The quantity of documents on each of these websites varies. "GEOJOURNAL" is the most accessed source out of the ten documents. Out of the 181 sources used, the least common categories are "ENVIRONMENTAL, DEVELOPMENT and SUSTAINABILITY" (13 sources), "ENVIRONMENTAL SCIENCE AND POLLUTION RESEARCH" (11 sources), and "INTERNATIONAL JOURNAL OF SUSTAINABILITY DEVELOPMENT A" (28 sources each).

Figure 3. Most relevant sources

The papers with the highest amount of citations worldwide are LIU X, 2020, NATURE SASTAIN (558 citations), HALL DM, 2017, CONSERV BIOL (428 citations), WANG J, 2019, SCI TOTAL ENVIRON & YU C, 2018, TRANSP RES PART B METHODOL (324 & 317 citations, respectively). With 136 citations, WOODRUFF SC, 2016, LANDSC URBAN PLANN is the least referenced document.

Figure 4. Most cited documents

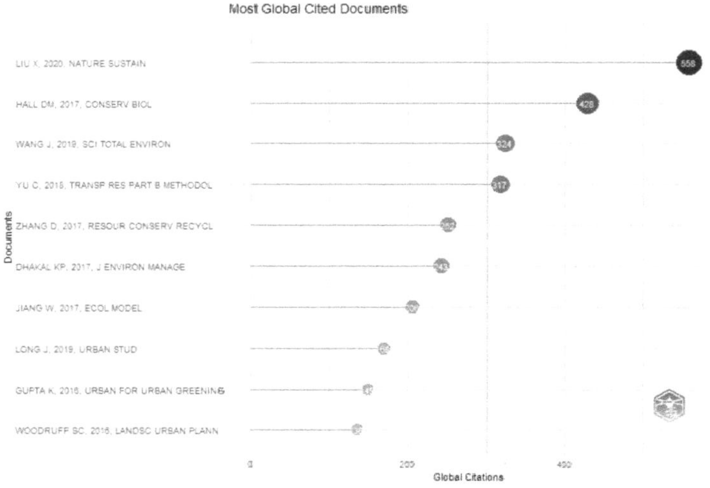

- **Bradford's Law**

The diagram shows how Bradford's Law may be used to find important sources in the subject of urban and environmental studies. The "core sources" zone is formed by a limited number of publications, including GeoJournal, Environment and Development, and Environmental Science and Policy, which produce a disproportionately large number of articles. The number of articles per journal sharply decreases as we proceed along the x-axis, which represents ranking sources. This distribution shows that the majority of pertinent material is concentrated in a small number of important journals. This information is helpful for academics who want to efficiently focus their literature reviews as well as for organizations who want to maximize their journal subscriptions for thorough subject coverage.

Figure 5. Core sources

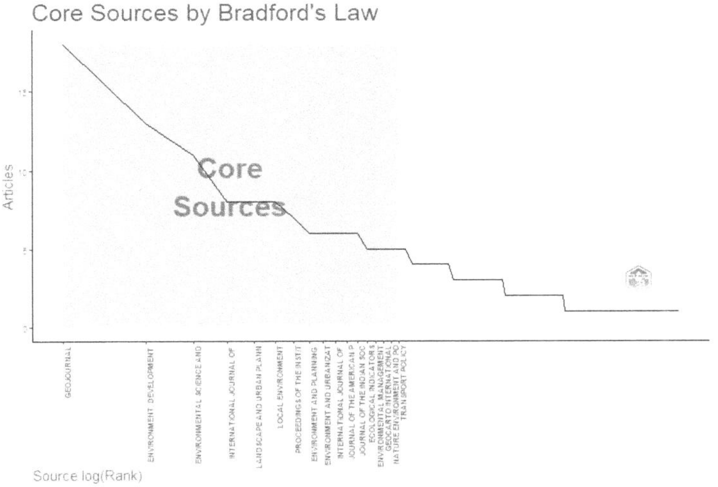

- **Sources Production Over Time**

A line graph comparing the quantity of documents from six distinct sources is shown in Figure 6. The X and Y axes stand for the years 2016 through 2025 and the amount of paper, respectively, while each source is represented by several colorful lines. The sources, which include IEEE, Access, Sensors, and others, are identified by unique colors such as green, blue, yellow, and red. The overall growth in documents for all resources throughout time is depicted in these graphs. An analytical tool for comprehending head publishing partners' perspectives on sustainability and urban planning is Figure 6. The sources are as follows: "Local Environment," "Environmental Science and Pollution Research," "Environmental Development and Sustainability," "Landscape and Urban Planning," "International Journal of Urban Sustainable Development," and "Environmental Science and Sustainability."

Figure 6. Source the production on time

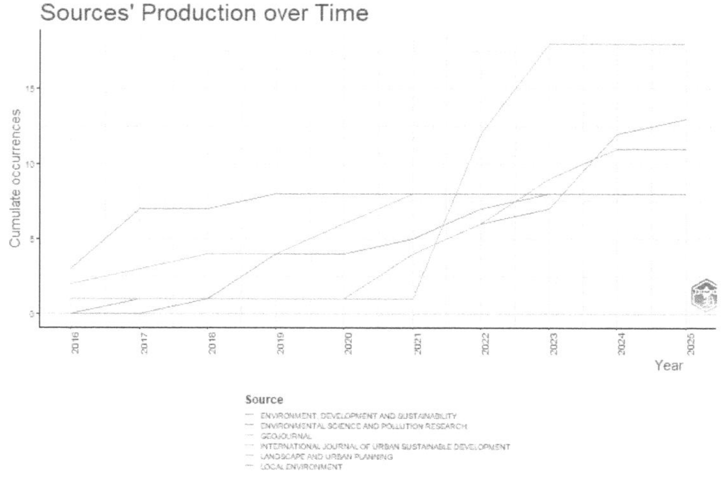

- **Most Relevant Author**

A line graph of the authors who have authored the most pertinent articles and their total number of documents is displayed in Figure 7. The link between two variables is represented by points in a particular kind of graph known as a point graph. The Y-axis shows the authors, while the X-axis shows the total number of documents. Data scales and classifications are depicted using axes, which are lines. They might be horizontal or vertical. The locations of the data points are shown by the blue dots. The values seen in the data points indicate the outcomes of measurements or observations of variables. The number of citations the author included from each source is shown in Figure 7. The light blue circle indicates the least relevant citation, while the dark blue circles indicate the most relevant citation.

Based on the graph, DESLATTE A & LI Y are the most prolific authors in the subject of sustainability and urban planning, with 10 citations each; LI Y, RUDRA S, WANGX, and WU J rank second with four citations each. The most significant studies in the fields of urban planning and sustainability's culture of ongoing learning have been authored by these scholars. It appears that they have made significant contributions to the discipline because other scholars have extensively acknowledged the work. The graph also illustrates the current surge in studies on urban planning and the culture of continuous sustainability. The rise in citations for publications published in the last 12 months is evidence of this.

Figure 7. Most relevant authors

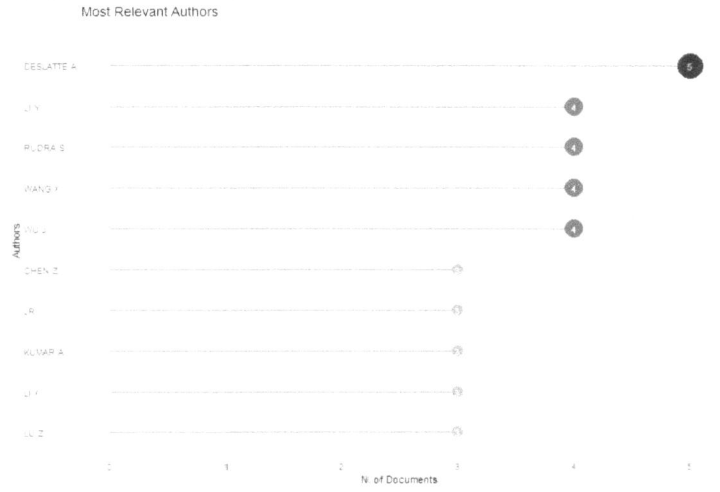

Although the majority of contributors have a similar impact level, indicating a consistent contribution in terms of citations or academic significance, it is clear from figure 8 that their sentiment ratings differ. Divergent opinions on tone, impact, or community reception are reflected in this variance. Two contributors in particular stand out for having substantially higher sentiment scores, which suggests a more positive response or connection with the principles of sustainable urban development. Because of their research's relevance, accessibility, or progressive character, these outliers point to thought leaders whose work more strongly aligns with contemporary urban sustainability narratives.

Figure 8. Authors local impact by H index

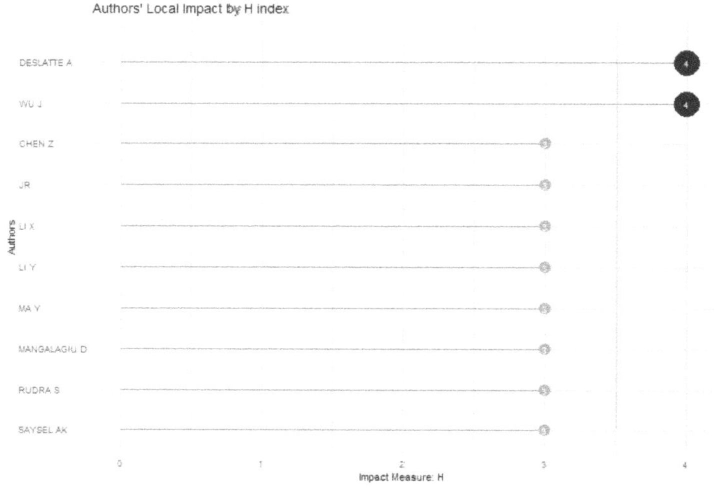

- **Author's Production Over Time**

The chart titled "Authors' Production over Time" provides an overview of the research output of various authors from 2016 to 2025. Each line shows the publishing history of each author, with circles representing the number of articles (N.Articles) created in each year. More citations are indicated by darker colors. The diameters of the circles, which vary from 1 to 2, indicate the quantity of articles and the depth of color, which indicates the total citations (TC) for the year.

Significant work by DESLATTE A is shown by comparatively larger circles around 2017 and 2024, respectively, indicating greater research effort and impact in those years. In contrast to author LU Z, whose efforts were scattered but nevertheless received attention, authors LI Y, RUDRA S, WANGX, and WU J have continuously contributed throughout time, despite writing fewer works and having a lower citation effect. This graphic provides insight into each author's output and influence throughout time.

DESLATTE The darker color denotes the total amount of citations received annually; a writer has previously received more citations. The size of the circles indicates the quantity of papers produced. According to the statistics, certain authors appear to have stayed or gained more clout as urban planning has become more prevalent in international business, which has helped the field of sustainability research advance and improve.

Figure 9. Authors' productivity over time

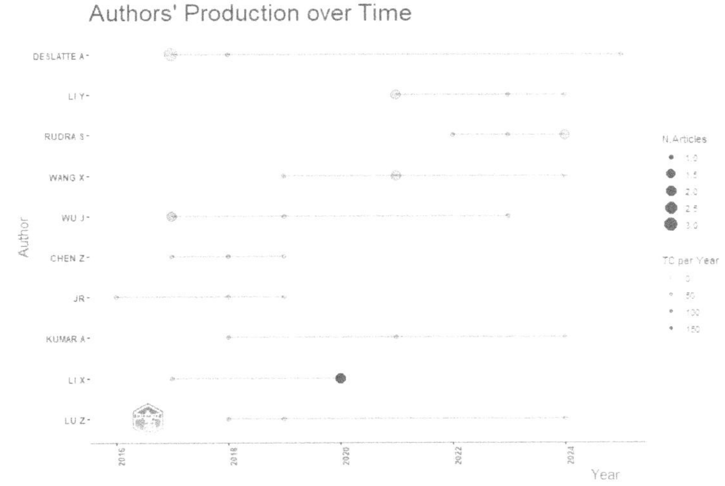

- **Three-Field Plot**

The research methodology known as "bibliometrics" uses statistical analysis and quantitative techniques to find recurring themes in a corpus of books. This is an excellent tool, particularly for mapping projects where the lights may be useful. Regarding the many, diverse, complex, and often controversial fields of research. Based on Lotka's Law, Biblioshiny, an online Bibliometrix interface, creates three-dimensional depictions of writers' productivity. Furthermore, Biblioshiny is used to create a word cloud based on the most frequently used word keys.

The information flowing between the sources (left field), publishers, authors, and keywords (right field) may demonstrate a Biblioshiny three-field plot technique. In order to provide a more relevant representation, just a small number of items from each field are examined in this study. This approach offers a thorough summary of the relationships and frameworks present in the scientific literature for a certain topic.

As seen in Figure 10, the majority of the research's publications stress the significance of "URBAN PLANNING and SUSTAINABILITY." Alongside these articles are publications from journals such as "Environmental, Development and Sustainability," "Environmental Science and Pollution Research," "GEOJOURNAL," "International Journal of Urban Sustainable Development," "Local Environment," and "Landscape and Urban Planning. The authors with the most papers on the application of sustainability and urban planning are DESLATTE A, LI Y, RUDRA S, WANG X, and WU J.

Figure 10. Three-field plot

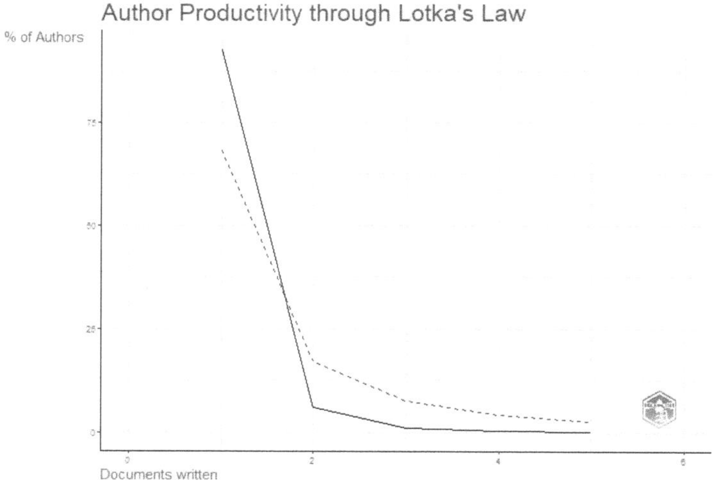

• **Most Relevant Affiliations**

Affiliates for a certain amount of time. The relevant contribution of each affiliate to the rise of document output is shown in this table. This table is important since it shows how each colleague helped create a certain document or work of literature. The most noteworthy or active connections related to the topic of inquiry may be found by using this type of investigation.

With 181 publications published in the last ten years, Figure 11 demonstrates the notable increase in the usage of sustainability and urban planning. With 26 papers, the most affiliations are found in the environmental, development, and sustainability programs at the University of Minnesota, Arizona State University, Beijing Normal University, and the University of California. The University of Maryland, Vidyasagar University, and Peking University follow with thirteen. East China Normal University has the fewest affiliations (10), followed by Texas A&M University of the United States and Aligarh Muslim University of Aligarh, Uttar Pradesh, with 11. This is because of the affiliate's significant contributions to the field's research on sustainability and urban planning at that era.

Figure 11. Most relevant affiliations

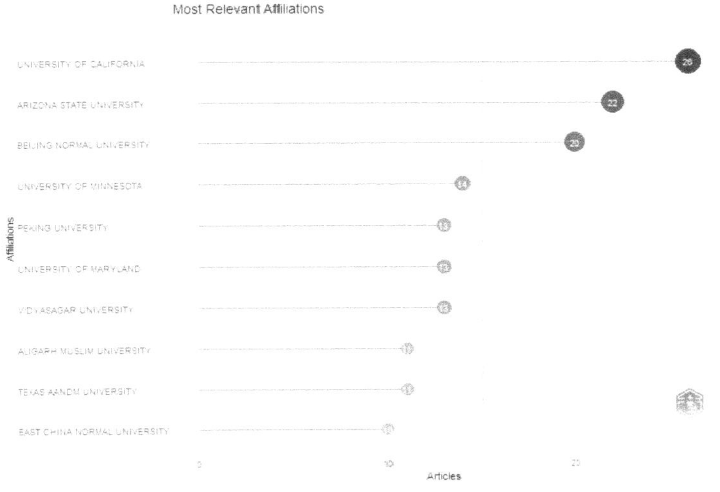

- **Affiliation's Production Over Time**

A longitudinal study of scholarly production pertaining to the topic Mapping Urban Planning and Sustainability is presented in the chart "Affiliations' Production over Time" (Figure 12). This analysis is based on the number of articles published by different academic institutions between 2016 and 2024.

The horizontal axis indicates the years, the vertical axis counts the number of publications published, and each colored line represents a different university. According to the data, Arizona State University has continuously led in terms of the amount of publications, indicating a steady level of scholarly interest in the topic. Fudan University and Beijing Normal University both have a notable rising tendency, suggesting an increase in research contributions in recent years.

Other universities like the University of California, the University of Maryland, and the University of Minnesota display more steady rises in their publishing output, suggesting either rising interest or a more defined niche within the wider field. A recent institutional drive toward urban sustainability research may be the reason for Vidyasagar University's notable late but substantial increase in publication count.

All things considered, the graph illustrates the development and comparative output of significant academic associations, stressing the dynamics of worldwide research as well as the growing scholarly attention to sustainable urban planning in recent years.

Figure 12. Affiliation production over time

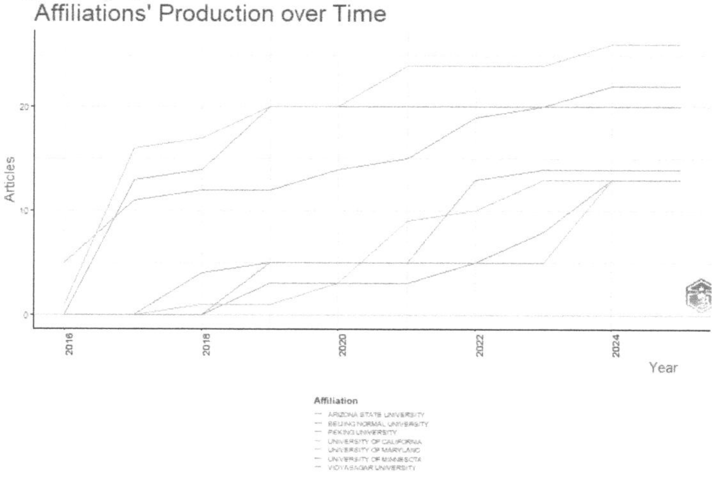

- **Corresponding Author's Countries**

Figure 13's claim that India has the most SCP and MCP is supported by the fact that the majority of the corresponding writers are from industrialized countries. The United States, China, Australia, the United Kingdom, Canada, Germany, Georgia, Iran, Korea, the Netherlands, Brazil, Ethiopia, France, Hong Kong, Malaysia, Mexico, Norway, Portugal, and Sweden are among the other countries that collaborate. This trend illustrates the significance of academic and research institutions in these fields, which are usually at the forefront of research on urban planning and sustainability. Nonetheless, emerging economies have made a notable contribution, particularly those in Asia, where countries like China, India, and the United States are producing important research at a quick pace. The representation of numerous geographical areas highlights the significance of transformational urban planning and ongoing learning in a variety of sustainability and economic circumstances.

Figure 13. Corresponding countries

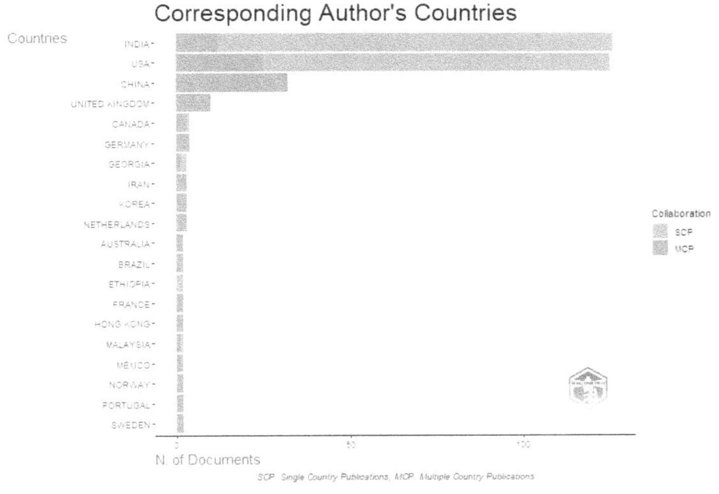

- **Top 5 Countries for Production Over Time**

Figure 14 of the chart "Country Production over Time" shows the research production of the top 5 nations in the field of sustainability and urban planning between 2016 and 2025. In terms of publishing volume, the US leads constantly, and by 2025, it will have about 500 articles. China and India come next, with India showing a notable surge in production following 2021, indicating a rising emphasis on urban sustainability at the national level. China's contribution is stable yet stabilizing, as seen by its steady increase until plateauing at the conclusion of the era.

In contrast, the USA, UK, and Germany continue to increase moderately and steadily, indicating a longer but less aggressive research trajectory. These trends show how the global research environment is changing, with emerging nations like India quickly gaining academic recognition in sustainability and urban planning studies while the United States continues to contribute the most.

Figure 14. Top 5 countries for production

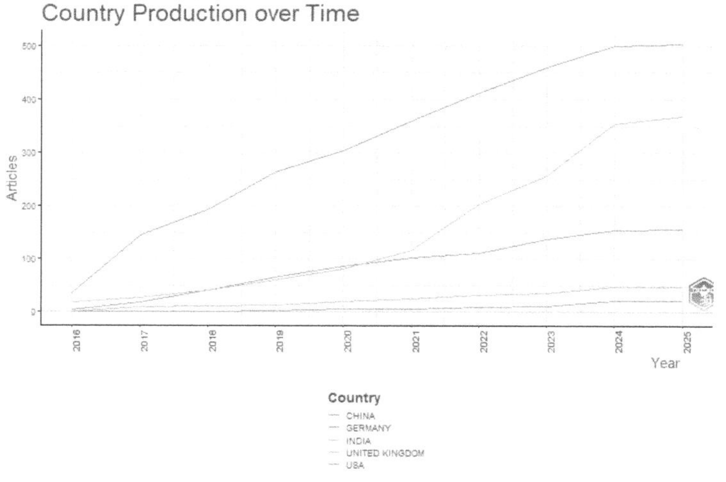

- **Most Cited Countries**

Based on two metrics—the quantity of citations and published documents—figure 15 graphically depicts the scholarly influence of top countries in the field of urban planning and sustainability. Each bubble represents a nation, and the vertical axis shows the average number of citations per document, while the horizontal axis shows the overall number of citations. Each bubble's size corresponds to the total number of articles, providing a thorough understanding of effect and output.

Due to its large number of publications, high average citation rate, and largest number of citations, the United States is clearly the most influential country in this field of study. In terms of volume, China and India are likewise quite significant, but their average number of citations per document is lower, suggesting a wider rather than deeper citation footprint. Other nations that strike a balance between average impact and publishing production include the United Kingdom and Germany. The productivity of nations is clearly displayed in the graphic, along with the relative caliber or impact of their research contributions to the conversation on urban sustainability.

Figure 15. Most cited countries

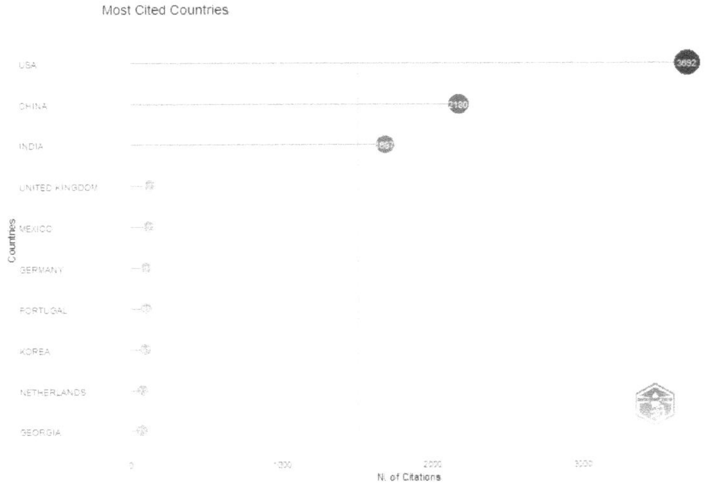

- **Most Relevant Words**

Based on the bibliometric analysis of "Mapping urban planning and sustainability," the term "Sustainable Development" appears most frequently in the literature. This prominence demonstrates the importance of urban planning in creating business cultures that embrace flexibility and continuous learning. As seen in figure 16, the term "sustainable development" was used 233 times, frequently in conjunction with other significant phrases like "urban planning," "sustainability," and "urbanisation," highlighting its critical significance in fostering organisational transformation. 178, 120, and 110 were other significant terms that were employed in conjunction with it. Given how often this phrase is used, it implies that urban planning is crucial for establishing corporate cultures where sustainability is woven throughout the organisation and for serving as a topic of discussion when discussing organisational expansion.

Figure 16. Top relevant words

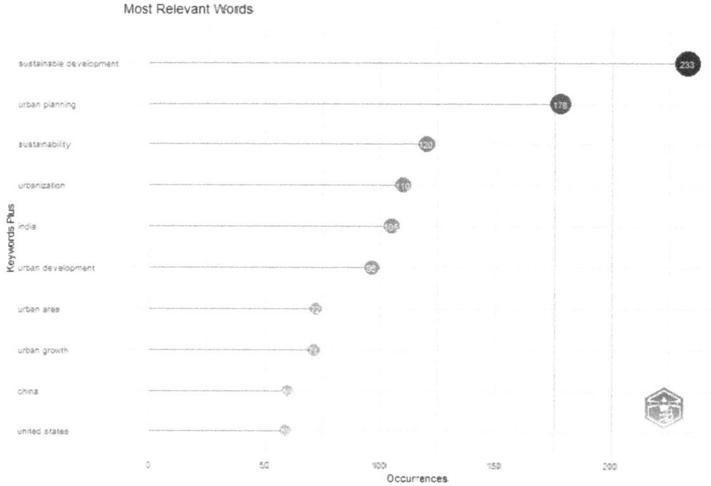

- **Word Cloud of Titles using Biblioshiny App.**

This picture is a word cloud that depicts concepts related to sustainability and urban planning. An illustration of the frequency or significance of a word in a text is called a word cloud. Larger words convey greater pressure or importance than comparable ones, and their size indicates how frequently or how significant they are.

Figure 17. Word cloud

The Biblioshiny Bibliometrix visualisation in the Figure 17 package indicates that the keywords for sustainable learning and urban planning in environmental studies will exhibit a consistent weekly trend between 2016 and 2025. When used visually, keywords or phrases with more volume and density appear in a bolder, larger typeface. As a result, we may see more information about the pattern or focus of keywords by using these terms in research publications.

- **Tree Map**

The frequency and distribution of keywords linked to scholarly works on urban planning and sustainability are depicted in the tree map. A keyword is represented by each block, whose size is determined by how frequently it occurs in the dataset. With 233 instances (11%), "sustainable development" stands out as the most prevalent term on the list, demonstrating its importance and frequent appearance in studies on the topic. The significant utilisation of other well-known phrases like "urban planning" (178), "sustainability" (120), and "urban development" (96) further emphasises the thematic focus of the academic debate.

The geographical focus of most of the literature is also reflected in the remarkable prevalence of country-specific phrases like "India," "China," and "United States." Keywords that integrate environmental, technological, and governance issues, such as "urbanisation," "land use," "climate change," "decision making," and "GIS," imply a multidisciplinary junction. Niche phrases like "greenspace," "policy making," and "ecosystem services" indicate that people are becoming more interested in the ecological and local aspects of urban sustainability. All things considered, the tree-map provides a thorough overview of the field's research trends and thematic areas of focus.

Figure 18. Tree map

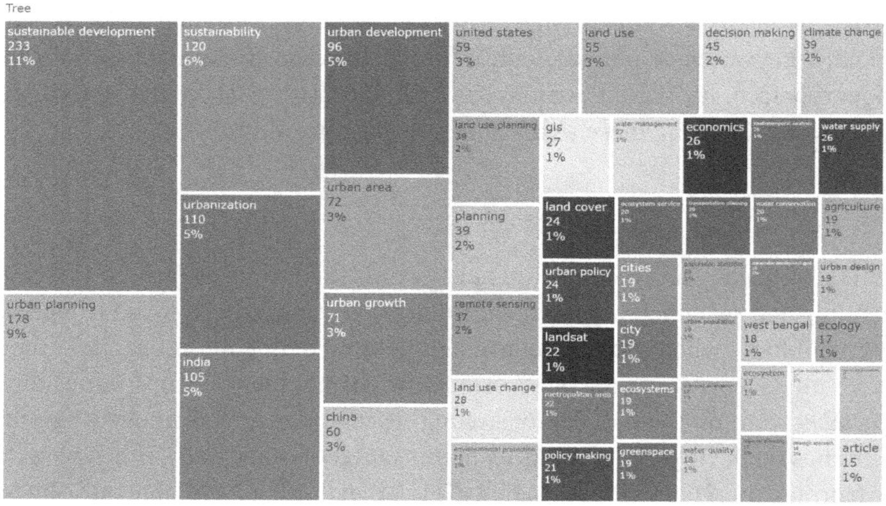

- **Word's Frequency Over Time**

The line graph shows how important terms used in urban planning and sustainability literature changed over time between 2016 and 2025. With a steady increasing trend and more than 200 total references by 2025, "China" stands out as the most commonly occurring term among all those examined. This implies that China will be a major area of study in the context of sustainable urban development. Next in line is "India," which also shows a noticeable rise in mentions, especially after 2020, suggesting that academic interest in India's urbanisation and sustainability initiatives is expanding.

Other often used phrases include "sustainability," "sustainable development," and "urban planning," all of which have grown steadily over time, underscoring their fundamental position in academic discourse. Similar trends may be seen in terms like "urban growth," "urbanisation," and "urban development," which highlight the field's emphasis on the dynamics of growing cities. Furthermore, phrases like "United States" and "urban area" are frequently used, indicating that the conversation is global in scope. The general upward trend for all phrases shows that interest in the topic is expanding, and more research is being done on various aspects of urban sustainability.

Figure 19. Word's frequency

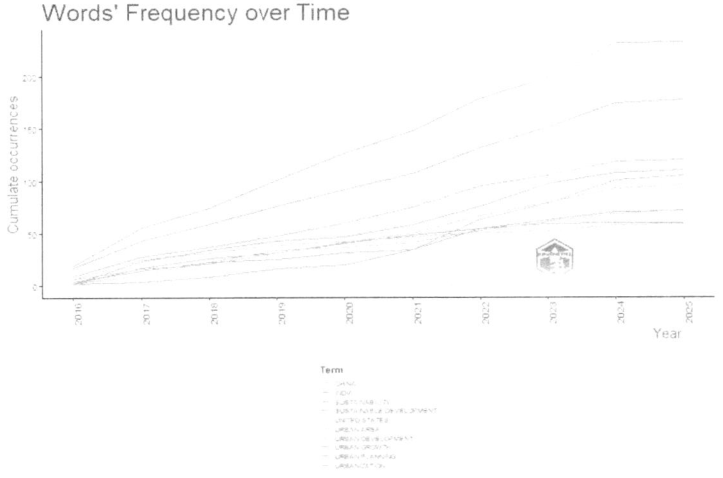

- **Trends Topics**

The birth and development of key phrases in the field of urban planning and sustainability research from 2016 to 2025 are graphically represented in the "Trend Topics" chart (figure 20). The size of the bubble indicates how frequently a term appears in academic journals, and each term is plotted according to the year it became well-known. Notably, phrases like "sustainable development," "urban planning," and "sustainability" have maintained their importance across time, indicating their fundamental relevance in the conversation. The terms "urban growth," "urban development," and "decision making" became popular, especially between 2019 and 2021, when there was a greater worry worldwide about the need for adaptive planning techniques and the rate of urbanisation.

In the years after 2020, more current entries like "solar energy," "vegetation cover," and "COVID-19" stand out as examples of new research topics brought on by environmental changes and the worldwide epidemic. Furthermore, national terminology like "China," "India," and "Malaysia" highlight the expanding regional interest in and contributions to the global sustainability discourse. Consistent focal areas are indicated by the appearance of other themes such as "water management," "environmental protection," and "public policy" over several years. The variety of words and their fluctuating usage highlight how dynamic and complex urban sustainability research is, with shifting objectives impacted by both local and global events.

Figure 20. Trendy topics

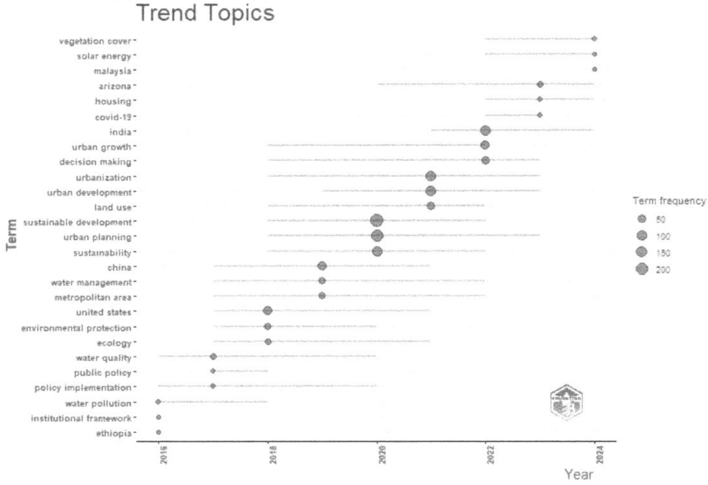

- **Co-Word Net**

Vosviewer offers three main types of visualisation: overlays, network visualisation, and density visualisation. Overlays show traces of earlier research on a certain topic or area. To create these overlays, "co-citation analysis" is the technique employed. Two items are regarded as co-closely associated papers when they are frequently discussed jointly in another document. Figure 21 illustrates how research on sustainability education and urban planning has expanded dramatically during the past few years.

Figure 21. Word net

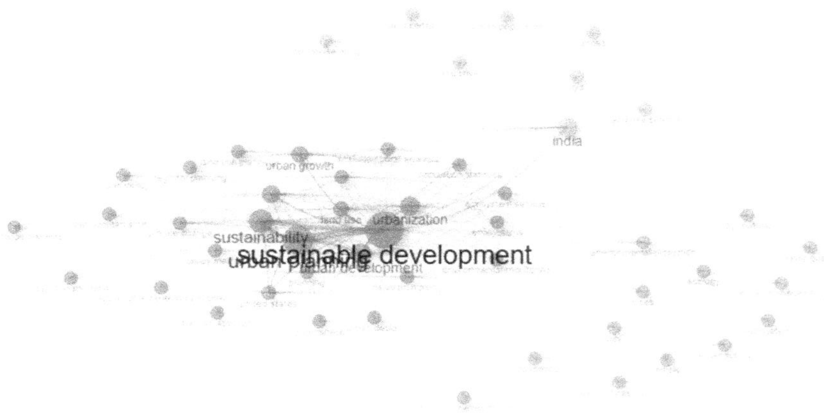

This has been shown in numerous studies using various terms. Here are a few recent advancements in sustainability and urban planning. Numerous studies that employ words like data analytics and the digital transformation sector make this clear. 4.0. Corporate awareness and responsibility are driving an increase in research on the role of urban planning and a sustainable culture in tackling global concerns, such as social, economic, and climatic issues. This is shown by a number of studies that use terms like sustainability, environmental management, and economic and social implications.

Figure 22. Word net node degree

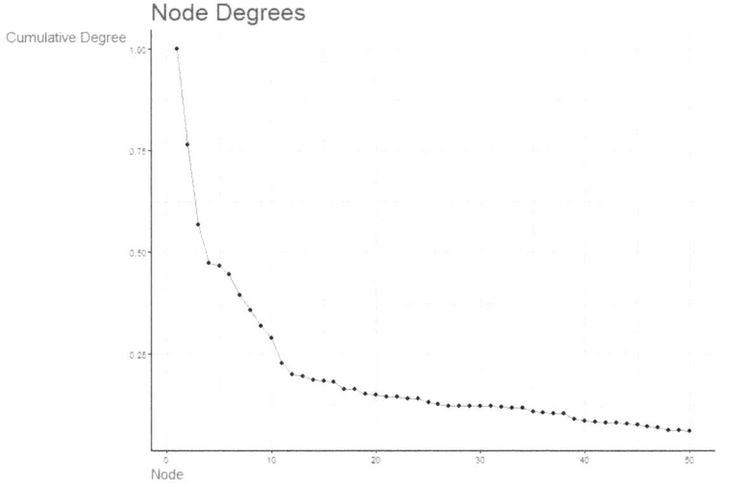

- **Thematic Map**

Figures 23 and 24 are a thematic map that illustrates the bibliometric analysis of the field of study on "Mapping Urban Planning and Sustainability." Two dimensions—development degree (density) and relevance degree (centrality)—are used to divide the chart into four quadrants. "Sustainable development," "urban planning," and "sustainability" are shown in the upper-right quadrant, referred to as the "Motor Themes," which denotes subjects that are both well developed and essential to the field. This reflects their basic and interrelated roles within the research domain. The upper-left quadrant, titled "Niche Themes," on the other hand, has the terms "economics," "environmental protection," and "water management," indicating that these fields are highly developed but have little sway over the remainder of the field.

Figure 23. Thematic map

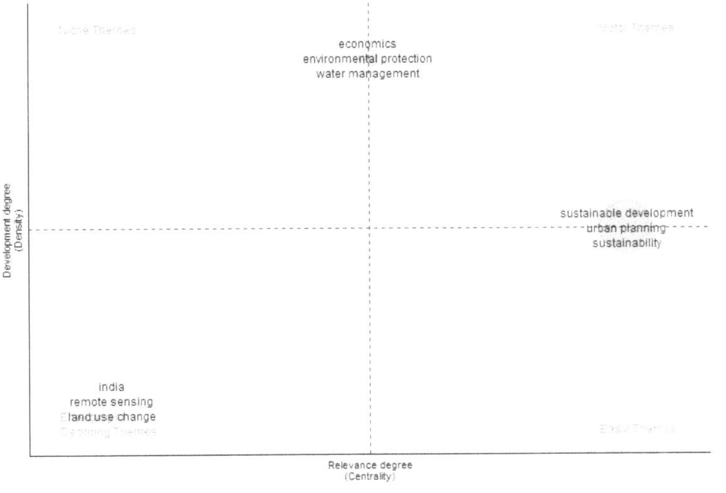

With no major themes shown other than a slight extension of sustainability notions, the "Basic Themes" (lower-right) in the lower quadrants highlight topics that are significant to the discipline but less internally developed. under contrast, topics like "India," "remote sensing," and "land use change" under the "Declining or Emerging Themes" quadrant (bottom-left) currently exhibit lower centrality and development, which may indicate either emerging fields of study or areas of declining research interest. The general layout of the map indicates areas for future research into more specialised or emerging subjects while emphasising sustainability and urban planning as key pillars.

Figure 24. Co-word net

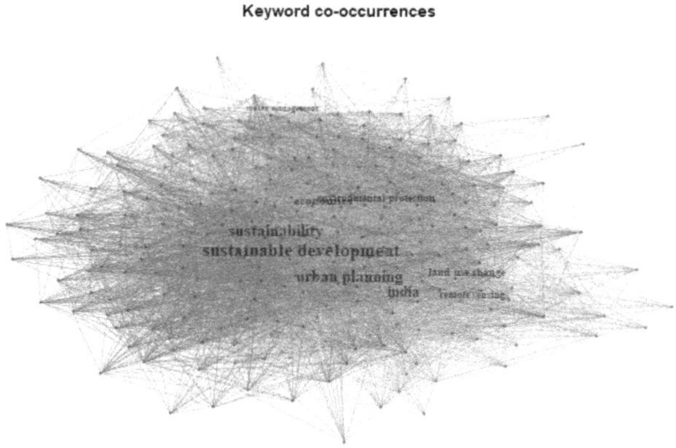

- **Collab Net**

The co-authorship network for the topic "Mapping Urban Planning and Sustainability," which highlights partnerships amongst scholars in the field, is depicted in Figures 25 and 26. Individual authors are represented by each node, while co-authored works are indicated by the links connecting them. Each node's size corresponds to the author's influence or output of publications within the network. "Mav Sayselek" stands out as the most notable individual, indicating a pivotal position in research output and collaboration concerning sustainability and urban planning.

Figure 25. Collaboration net

Different hues represent different groupings or clusters of writers who work together more often than not but less often with other authors. Diverse study communities or topic focusses within the larger subject are indicated by this division into clusters. While some groups seem more isolated, indicating either autonomous research tracks or new researchers, others are tightly knit with many internal links. All things considered, the map sheds light on the collaborative framework and significant figures influencing this field of study.

Figure 26. Collaboration node degree

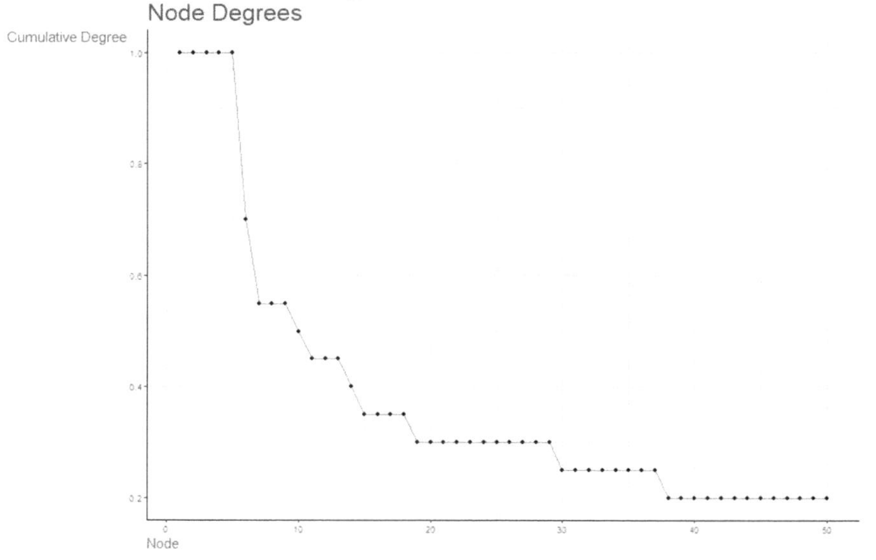

DISCUSSION

The research landscape is dynamic and changing, according to the bibliometric analysis of urban planning and sustainability. Scholarly interest in incorporating sustainable development principles into urban settings has grown significantly during the last few decades. This increase is a result of people becoming more conscious of environmental issues like resource depletion, climate change, and urbanization-related socioeconomic inequality. It is clear from charting the field's intellectual structure that sustainability in urban planning is not a single, independent subject but rather the nexus of several academic fields, such as environmental science, engineering, social sciences, and economics.

The grouping of study topics around important ideas like resilience, smart cities, green infrastructure, sustainable urban development, and governance is one of the analysis's important conclusions. These clusters point to a field that is becoming more diverse as sustainability objectives are being incorporated into the redefining of basic urban planning ideas. For example, the emergence of smart cities emphasises how technology and data-driven solutions are essential to creating sustainable urban systems. A move towards a more proactive planning philosophy is demonstrated by the emphasis on resilience, which tackles the need for cities to adjust to both environmental and social shocks.

According to the chronological analysis of publications, there has been a noticeable increase in interest in sustainable urban planning since the early 2000s. This trend is probably due to significant international initiatives like the Sustainable Development Goals (SDGs) and the Millennium Development Goals (MDGs) of the United Nations. In particular, it appears that the SDGs' emphasis on sustainable cities (Goal 11) served as a catalyst, spurring a surge of studies on inclusive, secure, resilient, and sustainable urban environments.

Regarding geographic contributions, the report emphasises that the majority of research production is centred in developed regions, particularly in North America, Europe, and parts of Asia. Major contributions include the United States, China, the United Kingdom, and Australia. Disparities in research funding, institutional capabilities, and governmental agendas are reflected in this geographic bias. It also indicates the need for more extensive international involvement, especially from quickly urbanising regions like South Asia and Africa, where sustainability issues are arguably the most pressing.

Attention should also be paid to the patterns of collaboration in the field. The increasing number of multinational and multi-author collaborations is a noteworthy trend. This suggests that urban sustainability is acknowledged as a multifaceted, international problem that cuts across national borders. The progress of the area is frequently enhanced by collaborative networks, which increase the diversity of viewpoints, approaches, and solutions. However, the bibliometric results also suggest the existence of silos, in which particular groups of scholars maintain a high degree of isolation, which may hinder interdisciplinary creativity.

Citation analysis clarifies key publications and underlying beliefs that have influenced the conversation. The recurring citation of seminal research pertaining to ideas such as participatory planning, compact cities, and ecological urbanism highlights their ongoing significance. Furthermore, the existence of critical viewpoints challenging the dominant sustainability narratives suggests a robust academic discourse, which promotes the development of more complex and situation-specific methodologies.

Lastly, new patterns that will probably shape the direction of urban sustainability research are revealed using keyword analysis. There has been a considerable increase in the use of terms like "climate adaptation," "urban metabolism," and "nature-based solutions," indicating a shift away from piecemeal interventions and towards systemic, integrated measures. These patterns suggest that the industry is evolving and shifting away from theoretical frameworks and towards quantifiable, actionable results.

All things considered, the bibliometric analysis offers a useful road map for comprehending the development, dynamics, and potential paths of sustainability and urban planning research. It draws attention to both successes and shortcomings, providing important information for academics, decision-makers, and professionals who are dedicated to creating more sustainable urban futures.

CONCLUSION

The bibliometric analysis of sustainability and urban planning demonstrates how the area has significantly changed and expanded in the last few years. Sustainability has become a crucial element of urban planning discussions and practices as cities around the world deal with increasing environmental, social, and economic constraints. Research production has been steadily increasing, according to the data, and multidisciplinary partnerships that combine expertise from disciplines like environmental science, economics, engineering, and social policy have significantly increased. This pattern emphasises how complicated sustainable urban development is by nature, necessitating multi-sectoral and integrative methods.

Importantly, theme cluster mapping shows that urban planning's understanding of sustainability has advanced beyond its initial conceptions. In order to consider concerns like equity, smart technologies, climate resilience, and participatory governance, traditional ideas of urban planning have been reinterpreted. Keywords like "resilience," "smart cities," "urban green spaces," and "climate adaptation" are frequently used, which points to a trend in research towards practical, solution-focused studies that aim to close the gap between theory and practical application. Additionally, it seems that the impact of global policy frameworks—most notably the Sustainable Development Goals—has inspired academic and professional communities to work towards common sustainability goals.

The report does, however, also highlight certain enduring difficulties. There is still a significant concentration of research in the Global North, suggesting a geographic imbalance that would restrict the applicability of results to the Global South's increasingly urbanising areas. These areas frequently deal with unique and pressing sustainability issues that are not well covered in the literature, such as informal settlements, inadequate infrastructure, and governance constraints. Furthermore, even while interdisciplinary cooperation is expanding, more needs to be done to incorporate other viewpoints, especially those of under-represented groups whose opinions are vital to inclusive urban sustainability.

The comparatively dispersed nature of the research landscape is another noteworthy finding. Even while certain academic networks show a lot of links, there are still isolated groups that can profit from more extensive exchange of concepts and approaches. The growing number of critical and reflective studies being cited is encouraging because it shows that the field is self-aware and prepared to challenge conventional wisdom, which is a prerequisite for creating more practical and context-sensitive urban sustainability strategies.

To sum up, the bibliometric mapping provides a thorough summary of the growth and diversity of research in urban planning and sustainability. It draws attention to the field's increasing complexity, adaptability to global sustainability agendas, and steady progress towards more inclusive and useful frameworks. However, it also emphasises how critical it is to close current gaps, especially with regard to interdisciplinary integration and regional representation. In order to make sure that urban sustainability programs are fair, contextually appropriate, and equipped to handle the intricate problems of urbanisation in the twenty-first century, researchers, legislators, and practitioners must cooperate going forward.

Future Implications

Building on the bibliometric study's findings, a number of significant ramifications for next studies and applications in sustainability and urban planning can be found. First, there is a pressing need to broaden the scope of research activities. Because the challenges of urban expansion, inequality, and environmental degradation are most severe in rapidly urbanising parts of Africa, South Asia, and Latin America, scholars must pay more attention to these areas. In order to generate knowledge that is more internationally representative and to produce solutions that can be adjusted to different socio-economic, cultural, and environmental contexts, future research should aim to incorporate a variety of urban scenarios.

Second, it will be crucial to place more focus on transdisciplinary and interdisciplinary research. Integration of technical, social, economic, and environmental viewpoints is necessary to address complex urban sustainability issues. Future studies should encourage interdisciplinary cooperation and involve stakeholders outside of academia, such as practitioners, policymakers, and community people. Co-produced knowledge that combines local knowledge with scientific competence can result in more contextually aware and successful urban solutions.

The necessity of a more thorough examination of social justice and equality concerns within the sustainability agenda is another significant effect. The socio-spatial disparities that define many cities must be addressed in future urban planning studies to prevent sustainable solutions from unintentionally making already-existing inequities worse. Instead of being ancillary issues, ideas like environmental justice, the right to the city, and inclusive governance ought to be at the heart of sustainability research frameworks.

REFERENCES

Asif, M., Ersoy, A., Lodhi, R. N., & Cobanoglu, C. (2025). Mapping the metaverse-sustainability nexus in hospitality and tourism: A bibliometric approach. *Online Information Review*, *49*(3), 600–622. DOI: 10.1108/OIR-09-2023-0510

Çelik, K. T., & Şekeroğlu, A. (2023). The importance of geographical information systems in urban and landscape planning: A bibliometric analysis. *Megaron, 18*(4).

Chen, Y., Wang, Y., & Zhou, D. (2021). Knowledge map of urban morphology and thermal comfort: A bibliometric analysis based on citespace. *Buildings (Basel, Switzerland)*, *11*(10), 427. DOI: 10.3390/buildings11100427

Cicea, C. (2020). Mapping the research agenda of energy in urban areas: A bibliometric analysis. *Theoretical and Empirical Researches in Urban Management*, *15*(2), 66–80.

Cui, X. (2018). How can cities support sustainability: A bibliometric analysis of urban metabolism. *Ecological Indicators*, *93*, 704–717. DOI: 10.1016/j.ecolind.2018.05.056

D'Alpaos, C., & Andreolli, F. (2020). Urban quality in the city of the future: A bibliometric multicriteria assessment model. *Ecological Indicators*, *117*, 106575. DOI: 10.1016/j.ecolind.2020.106575

Fu, Y., & Zhang, X. (2017). Trajectory of urban sustainability concepts: A 35-year bibliometric analysis. *Cities (London, England)*, *60*, 113–123. DOI: 10.1016/j.cities.2016.08.003

Guo, P., Li, Q., Guo, H., Li, H., & Yang, L. (2021). A bibliometric and visual analysis of global urban resilience research in 2011–2020: Development and hotspots. *Sustainability (Basel)*, *14*(1), 229. DOI: 10.3390/su14010229

Hamman, P. (2017). *Definitions and redefinitions of urban sustainability: A bibliometric approach* (Vol. 11). Environnement Urbain/Urban Environment.

Hocevar, M., & Bartol, T. (2021). Mapping urban tourism issues: Analysis of research perspectives through the lens of network visualization. *International Journal of Tourism Cities*, *7*(3), 818–844. DOI: 10.1108/IJTC-05-2020-0110

İnce, E. C. (2025). Mapping the Path to Sustainable Urban Mobility: A Bibliometric Analysis of Global Trends and Innovations in Transportation Research. *Sustainability (2071-1050), 17*(4).

Janik, A., Ryszko, A., & Szafraniec, M. (2020). Scientific landscape of smart and sustainable cities literature: A bibliometric analysis. *Sustainability (Basel)*, *12*(3), 779. DOI: 10.3390/su12030779

Jia, B., Chen, Y., & Wu, J. (2021). Bibliometric analysis and research trend forecast of healthy urban planning for 40 years (1981–2020). *International Journal of Environmental Research and Public Health*, *18*(18), 9444. DOI: 10.3390/ijerph18189444 PMID: 34574368

Krabokoukis, T., & Polyzos, S. (2023). A bibliometric analysis of integrating tourism development into urban planning. *Sustainability (Basel)*, *15*(20), 14886. DOI: 10.3390/su152014886

Lazar, N., & Chithra, K. (2021). Comprehensive bibliometric mapping of publication trends in the development of Building Sustainability Assessment Systems. *Environment, Development and Sustainability*, *23*(4), 4899–4923. DOI: 10.1007/s10668-020-00796-w

Li, H., Xiang, Y., Yang, W., Lin, T., Xiao, Q., & Zhang, G. (2024). Green roof development knowledge map: A review of visual analysis using CiteSpace and VOSviewer. *Heliyon*, *10*(3), e24958. DOI: 10.1016/j.heliyon.2024.e24958 PMID: 38356566

Liang, D., De Jong, M., Schraven, D., & Wang, L. (2022). Mapping key features and dimensions of the inclusive city: A systematic bibliometric analysis and literature study. *International Journal of Sustainable Development and World Ecology*, *29*(1), 60–79. DOI: 10.1080/13504509.2021.1911873

Liao, P., Wan, Y., Tang, P., Wu, C., Hu, Y., & Zhang, S. (2019). Applying crowdsourcing techniques in urban planning: A bibliometric analysis of research and practice prospects. *Cities (London, England)*, *94*, 33–43. DOI: 10.1016/j.cities.2019.05.024

Lv, T., Wang, L., Xie, H., Zhang, X., & Zhang, Y. (2021). Exploring the global research trends of land use planning based on a bibliometric analysis: Current status and future prospects. *Land (Basel)*, *10*(3), 304. DOI: 10.3390/land10030304

Marvuglia, A., Havinga, L., Heidrich, O., Fonseca, J., Gaitani, N., & Reckien, D. (2020). Advances and challenges in assessing urban sustainability: An advanced bibliometric review. *Renewable & Sustainable Energy Reviews*, *124*, 109788. DOI: 10.1016/j.rser.2020.109788

Molaei Qelichi, M., Ghasemi, K., Tepe, E., & Murgante, B. (2025). Mapping Trends in Urban Livability Research: A Comprehensive Bibliometric Analysis. *Journal of Planning Literature*, •••, 08854122241312181. DOI: 10.1177/08854122241312181

Oliva, E. J. D., Sánchez-Torres, J. A., & Nino, J. S. S. (2022). City brand evolution research and future agenda: Bibliometric and science mapping analysis. *Heliyon*, *8*(8). PMID: 35991998

Pascariu, G. C., Banica, A., & Nijkamp, P. (2023). A meta-overview and bibliometric analysis of resilience in spatial planning–the relevance of place-based approaches. *Applied Spatial Analysis and Policy*, *16*(3), 1097–1127. DOI: 10.1007/s12061-022-09449-z PMID: 35495415

Pato, M. L., & Duque, A. S. (2025). Mapping Innovation and Sustainability in Rural Tourism: A Bibliometric Approach. *Sustainability (Basel)*, *17*(6), 2574. DOI: 10.3390/su17062574

Qian, H., Zhang, F., & Qiu, B. (2023). Deciphering the evolution, frontier, and knowledge clustering in sustainable city planning: A 60-year interdisciplinary review. *Sustainability (Basel)*, *15*(24), 16854. DOI: 10.3390/su152416854

Qu, S., Li, H., Wu, J., & Zhao, B. (2025). Environmental and Health Benefits: A Bibliometric and Knowledge Mapping Analysis of Research Progress. *Sustainability (Basel)*, *17*(5), 2269. DOI: 10.3390/su17052269

Razia, S., & Ah, S. H. B. A. B. (2022). Panoramic Mapping of Urban Social Sustainability: A 35-Year Bibliometric and Visualization Analysis. *Journal of Regional and City Planning*, *33*(2), 49–78. DOI: 10.5614/jpwk.2022.33.2.4

Sareen, S., & Haque, M. (2023). Exploring the Evolution and Trends in the Peri-Urban Planning: A Bibliometric Overview. *International Journal of Sustainable Development and Planning*, *18*(12), 3855–3866. DOI: 10.18280/ijsdp.181217

Sharifi, A. (2021). Urban sustainability assessment: An overview and bibliometric analysis. *Ecological Indicators*, *121*, 107102. DOI: 10.1016/j.ecolind.2020.107102

Sheng, S., Song, W., Lian, H., & Ning, L. (2022). Review of urban land management based on bibliometrics. *Land (Basel)*, *11*(11), 1968. DOI: 10.3390/land11111968

Siyavuş, A. E., & Aydın, T. N. (2022). A bibliometric analysis of urban sprawl. *Papers in Applied Geography*, *8*(2), 163–184. DOI: 10.1080/23754931.2021.1975307

Sokolova, M. V., Fath, B. D., Grande, U., Buonocore, E., & Franzese, P. P. (2024). The role of green infrastructure in providing urban ecosystem services: Insights from a bibliometric perspective. *Land (Basel)*, *13*(10), 1664. DOI: 10.3390/land13101664

Soliman, M., Cardoso, L., de Almeida, G. F., Araújo, A., & Vila, N. (2021). Mapping smart experiences in tourism: A bibliometric approach. *European Journal of Tourism Research*, *28*, 2809–2809. DOI: 10.54055/ejtr.v28i.2254

Van Winkle, T., Kotval-K, Z., Machemer, P., & Kotval, Z. (2022). Health and the urban environment: A bibliometric mapping of knowledge structure and trends. *Sustainability (Basel)*, *14*(19), 12320. DOI: 10.3390/su141912320

Wagner, N. (2019). Sustainability in Port cities–a bibliometric approach. *Transportation Research Procedia*, *39*, 587–596. DOI: 10.1016/j.trpro.2019.06.060

Wang, C., Lv, T., Cai, R., Xu, J., & Wang, L. (2022). Bibliometric analysis of multi-level perspective on sustainability transition research. *Sustainability (Basel)*, *14*(7), 4145. DOI: 10.3390/su14074145

Wang, H., Pan, Y., & Luo, X. (2019). Integration of BIM and GIS in sustainable built environment: A review and bibliometric analysis. *Automation in Construction*, *103*, 41–52. DOI: 10.1016/j.autcon.2019.03.005

Wang, M. H., Ho, Y. S., & Fu, H. Z. (2019). Global performance and development on sustainable city based on natural science and social science research: A bibliometric analysis. *The Science of the Total Environment*, *666*, 1245–1254. DOI: 10.1016/j.scitotenv.2019.02.139 PMID: 30970489

Yang, G., Zhang, P., Yu, F., & Zhu, X. (2023). A review on resilient cities research from the perspective of territorial spatial planning: A bibliometric analysis. *Frontiers in Ecology and Evolution*, *11*, 1300764. DOI: 10.3389/fevo.2023.1300764

Zaidi, A., Ajibade, S. S. M., Musa, M., & Bekun, F. V. (2023). New insights into the research landscape on the application of artificial intelligence in sustainable smart cities: A bibliometric mapping and network analysis approach. *International Journal of Energy Economics and Policy*, *13*(4), 287–299. DOI: 10.32479/ijeep.14683

Chapter 8

Health and Habitat: Rights of Sundarbans Tribals in Flood Risk Context

Oindrilla Ghosh
https://orcid.org/0000-0001-9609-0781
Christ University, India

Manjeri Subin Sunder Raj
https://orcid.org/0000-0002-2429-4395
Christ University, India

ABSTRACT

The Sundarbans region is home to diverse tribal communities facing significant health challenges exacerbated by flood hazards and wetland dynamics. Bridging health and habitat, this study explores the right to health for Sundarbans tribals through geostatistical insights. The correlation between environmental factors and health outcomes is critical, as these communities rely on wetlands for their livelihoods. Flooding not only disrupts access to essential resources but also increases vulnerability to waterborne diseases. By employing geostatistical methods, researchers can map flood-prone areas and wetland changes, facilitating proactive health interventions. This holistic approach underscores the need for policies that integrate environmental sustainability with healthcare access, ensuring that tribal populations can secure their right to health amidst changing climate conditions. Addressing these intertwined issues is vital for enhancing resilience and improving overall well-being in the Sundarbans.

DOI: 10.4018/979-8-3373-3206-2.ch008

INTRODUCTION

The Sundarbans, a UNESCO World Heritage Site, is a unique mangrove ecosystem located in the delta of the Ganges, Brahmaputra, and Meghna rivers (Das & Hazra, 2020). It spans parts of India and Bangladesh, supporting a rich tapestry of biodiversity and an extensive network of waterways (Sahana et al., 2021). However, this biological richness is often obscured by the ongoing human struggles for survival faced by the indigenous tribal communities (Kanjilal et al., 2010). The tribals in this region rely heavily on their natural habitat for sustenance, cultural identity, and economic well-being (Ahsan & Warner, 2014). Yet, they confront multifaceted threats, primarily amplified by climate change, including recurrent flooding and shifting wetland dynamics (Danda et al., 2019).

The interplay between health and habitat becomes critical in understanding the challenges these communities face (Panda et al., 2016). Health, as defined by the World Health Organization (WHO), is not merely the absence of disease but a state of complete physical, mental, and social well-being (Kanjilal et al., 2013). For the Sundarbans tribals, their health is intricately linked to environmental factors, such as access to clean water, food security, and exposure to flood hazards (Savacool, 2018). Seasonal floods, which are becoming increasingly erratic and intense due to climate change, undermine these basic human rights, leading to health crises ranging from waterborne diseases to malnutrition (Mistri, 2019).

Using geostatistical insights, researchers and policymakers can visualize and analyze the nuances of flooding risks and wetland dynamics (Alam & Bhardwaj, 2020). These insights can guide targeted interventions and policy decisions aimed at safeguarding the rights to health for the Sundarbans tribes (Roy, 2018). A more profound understanding of how habitat transformation influences the well-being of these communities is crucial, not just for their survival but also for preserving the unique ecological landscapes they inhabit (Rashad Al Mahamud et al., 2020). This exploration is imperative to raise awareness about the intersectionality of climate change, environmental justice, and health rights, advocating for more inclusive governance that respects and integrates indigenous knowledge and practices (Saif et al., 2016).

HISTORICAL EVOLUTION

The historical evolution of the Sundarbans region can be traced back to the merging of fluvial and tidal influences, creating a complex ecological landscape and supporting various tribal communities for centuries (Biswas, 2013). Historically, the Sundarbans had been a largely self-sustaining area with populations that were well-adapted to the cyclic nature of floods and seasonal changes (Pramanik et al., 2021). The tribals, who have lived in this unique ecosystem for generations, have developed intricate knowledge systems about local flora and fauna and sustainable practices that are essential for their livelihoods (Chowdhury, 2019).

Colonial rule in the 19th century drastically transformed the Sundarbans (Adams et al., 2020). The British aimed to exploit the region's vast natural resources, which resulted in the establishment of a plantation economy (Molinari, 2017). Unfortunately, this focus on agricultural expansion and resource extraction disregarded the environmental balance and traditional practices of local communities (Satterthwaite & Mitlin, 2014). The introduction of cash crops altered the hydrological dynamics of the area, making ecosystems more vulnerable to flooding and erosion (Satterthwaite, Huq, Reid, Pelling, & Romero Lankao, 2007). Moreover, the introduction of land ownership policies disrupted communal land practices and the traditional rights of the tribals (Soltesova, Brown, Dayal, & Dodman, 2014).

Post-independence, the focus shifted to development, but often at the expense of indigenous rights and environmental health (Ayers & Forsyth, 2009). Various government initiatives intended to bolster economic development in the Sundarbans, such as dam construction and urbanization, brought about significant habitat changes, further exposing tribals to the increasing threats posed by natural disasters (Cannon & Muller-Mahn, 2010). Flooding incidents became more severe, and loss of biodiversity jeopardized the traditional livelihoods of these communities (Twigg, 2007).

As concerns regarding climate change and environmental degradation have escalated in the 21st century, renewed attention has emerged around Indigenous rights and ecological conservation (Pervin et al., 2013). Advocacy groups, researchers, and policymakers have begun exploring innovative frameworks that integrate geostatistical methodologies to analyze flood risks and wetland dynamics (Olhoff & Schaer, 2010). By leveraging such insights, there is increasing recognition of the need to reclaim the voices of Sundarbans tribes in the discussions regarding their health outcomes, environmental rights, and future resilience against climate change impacts (Anguelovski & Carmin, 2011).

THEORETICAL FRAMEWORKS IN BRIDGING HEALTH AND HABITAT

The exploration of the right to health for Sundarbans tribals through geostatistical insights necessitates a robust theoretical framework that connects health, habitat, and environmental dynamics (Leck & Simon, 2012). Several theoretical paradigms can be applied to analyze these intersections, including the socio-ecological model, environmental justice theory, and human rights framework (Roy, Hulme, and Jahan, 2013).

Socio-Ecological Model: This model emphasizes the interplay between individual, relational, community, and societal factors in determining health outcomes (Anguelovski & Carmin, 2011). In the context of the Sundarbans, this framework can help elucidate how environmental factors such as flooding and wetland dynamics interact with social determinants like poverty, education, and access to healthcare (Tyler & Moench, 2012). By framing health concerns within this multi-layered approach, one can better understand how vulnerabilities among tribals are exacerbated by ecological changes, requiring an integrated strategy that encompasses environmental conservation and health promotion (Pelling, 2011).

Environmental Justice Theory: This framework addresses the disproportionate environmental burdens borne by marginalized communities, including tribal populations in the Sundarbans (Agrawal, Kononen, and Perrin, 2009). It underscores the importance of equitable distribution of environmental benefits and burdens and calls for the inclusion of affected communities in decision-making processes (Dodman and Mitlin, 2013). Through an environmental justice lens, we can critically analyze how flood risks and habitat degradation disproportionately impact Sundarbans tribals, exacerbating their already precarious health situations (Hurlimann & March, 2012). This theory would advocate for policies that ensure these communities have a voice in managing their habitats and health, promoting both resilience and equity (Few, Brown, and Tompkins, 2007).

Human Rights Framework: The right to health is enshrined in various international human rights instruments, asserting that health is not merely a commodity but a fundamental human right (Carmin et al., 2012). This framework can be powerful when applied to Sundarbans tribals, particularly in articulating their needs in the face of environmental challenges (Roberts et al., 2012). Integrating the right to health with ecological sustainability can illuminate pathways for advocacy, public policy, and community engagement efforts aimed at protecting tribal health in the face of climate change (Leck & Simon, 2012). This approach further necessitates that states recognize their obligations to ensure healthy habitats for all populations, affirming that the health rights of the Sundarbans tribals are respected, protected, and fulfilled (Almansi et al., 2014).

Incorporating geostatistics into these theoretical frameworks allows researchers to visualize and quantify the relationships between health and habitat dynamically (Syam, 2013). Geostatistical methods can provide evidence of correlations between flooding patterns, wetland health, and health outcomes within these communities (Brown, Dayal, & Rumbaitis del Rio, 2012). As researchers analyze how changes in the environment directly impact physical and mental health among Sundarbans tribals, they can leverage theoretical insights to recommend strategies that enhance resilience, advocate for rights, and ensure equitable health access (Corfee-Morlot et al., 2011).

USE OF AI AND ML IN BRIDGING HEALTH AND HABITAT IN SUNDARBANS

Artificial Intelligence (AI) and Machine Learning (ML) have transformative potential in addressing complex issues such as health and habitat in vulnerable regions like Sundarbans (Corfee-Morlot et al., 2011). The Sundarbans, which spans both India and Bangladesh, is a unique ecosystem characterized by its mangroves, waterways, and diverse tribal communities (Dewulf, 2013). The interface of health and habitat in this challenging environment presents unique challenges, particularly exacerbated by flood hazards and wetland dynamics (Norris, Stevens, Pfefferbaum, Wyche, & Pfefferbaum, 2008).

AI and ML can play critical roles in analyzing and interpreting complex geostatistical data related to flood risks and wetland conditions (O'Brien, 2012). For instance, predictive modeling utilizing machine learning algorithms can analyze historical weather patterns, rainfall data, and geographical information to forecast potential flooding events (Agrawal, 2010). By generating accurate predictive models, local authorities can better prepare for natural disasters, alerting communities to imminent dangers and thereby minimizing health risks associated with flooding, such as waterborne diseases and the resulting psychological impacts of displacement (Folke et al., 2005).

Moreover, AI-driven tools can be employed in health surveillance among the tribal populations of Sundarbans (Smit and Wandel, 2006). Real-time data collection and analysis through mobile health applications can track the prevalence of diseases related to environmental changes, such as malaria and cholera, which may rise during or after flooding episodes (Berkes and Ross, 2013). The integration of AI algorithms can aid in rapid diagnosis and treatment recommendations, ultimately enhancing healthcare delivery in remote areas (Adger et al., 2009).

In addition to direct health applications, AI and ML can significantly contribute to the sustainable management of wetlands (Berkes, 2009). Machine learning can analyze satellite imagery to assess vegetation cover, water quality, and biodiversity changes over time (Mertz et al., 2009). This information can be crucial for policymakers aiming to implement effective conservation strategies in the Sundarbans, which is influential in maintaining ecological balance and supporting the livelihoods of tribal communities (Eakin and Luers, 2006). Maintaining healthy wetlands can also mitigate flooding risks by enhancing natural barriers and absorbing excess water (Gómez-Baggethun et al., 2012).

Furthermore, AI and ML can facilitate participatory research and community engagement (Yohe and Tol, 2002). Tools that visualize complex data about health risks and environmental changes can empower tribal communities to make informed decisions about their health and habitat (Cutter et al., 2003). By fostering a collaborative framework where residents can view real-time data and understand its implications, they can adopt preventive health measures or advocate for necessary interventions from authorities (Armitage, 2005).

AI and ML's predictive capabilities can also influence socio-economic planning (Adger et al., 2005). For instance, integrating economic models with health outcomes can help identify vulnerable populations in the Sundarbans, allowing for targeted interventions that promote health equity (Biesbroek et al., 2013). By understanding the links between environmental health and human health outcomes, resource allocation can become more strategic and effective (Smit and Pilifosova, 2003).

ROLE OF AR, VR, AND MR IN BRIDGING HEALTH AND HABITAT IN SUNDARBANS

Augmented Reality (AR), Virtual Reality (VR), and Mixed Reality (MR) technologies possess the potential to revolutionize the approach to health and habitat issues (Berkes, 2009) in the Sundarbans, particularly for the tribal populations who face unique challenges due to flooding and wetland dynamics (Midgley, 1995). These immersive technologies can bridge the gap between digital data and real-world impacts, enhancing understanding, education, and community resilience (Elliot, 1993).

AR can provide on-the-ground insights by layering digital information over physical environments (Sovacool, 2011). For instance, local healthcare providers can use AR applications to visualize flood risk projections and health information directly in situ (Moser and Ekstrom, 2010). This application of AR can facilitate community workshops where tribal members interact with geostatistical data visually, enhancing their understanding of local health risks associated with flooding (Pelling, 2003). By incorporating AR into disaster preparedness training, communities can learn to identify safe zones (Agrawal, 2008), evacuation routes, and resource centers, fostering a proactive approach to health and safety (Pelling & Manuel-Navarrete, 2011).

VR offers an immersive experience that enables users to simulate various scenarios, such as flooding impacts on health and habitat (Mosher, 1979). For instance, VR can be used in educational settings to replicate flooding conditions and demonstrate the associated health risks and environmental changes (Omer, 1979). Such simulations can cultivate empathy and awareness among non-residents while training local communities on responding to flood hazards (Paiva, 1977). By experiencing these scenarios virtually, individuals can practice their responses to emergency situations without facing the real-life risks, thus improving their preparedness (Gray, 2006).

MR combines elements of both AR and VR, allowing users to interact with digital content in their real-world settings (Healy, 2008). For the Sundarbans tribals, MR can be utilized in healthcare training programs where community health workers practice diagnostic skills through simulations that provide real-time feedback (Peeters, 2012). This training is critical in regions where access to professional healthcare services may be limited (Erickson, 2012). Moreover, MR can assist in visualizing the impacts of health interventions on community wellbeing, fostering a culture of health literacy (Dominelli, 2010).

Additionally, the integration of AR, VR, and MR can enhance data visualization of geostatistical insights into flood hazards and wetland dynamics (Dubey et al., 2017). Access to dynamic modeling tools allows local residents to interact with predictions and scenarios tailored to their specific conditions, enabling informed community decision-making (Mishra et al., 2021). For example, sustainability initiatives can be illustrated through MR, where users can visualize potential outcomes of conservation efforts on health and habitat in their locality (Biswas & Nautiyal, 2020).

ADVANTAGES OF BRIDGING HEALTH AND HABITAT FOR SUNDARBANS TRIBALS

The concept of bridging health and habitat in the context of the Sundarbans tribals opens avenues for enhanced health outcomes, sustainable development, and resilience against environmental challenges (Rudra & Chattopadhyay, 2019). The region, known for its rich biodiversity and unique ecosystem formed by a complex network of wetlands, faces significant threats from climate change, flooding, and erosion (Rudra & Chattopadhyay, 2018). Integrating geostatistical insights into health assessments and habitat management can have several advantages for the tribal communities inhabiting these riverine and wetland landscapes (Satio et al., 2020).

Improved Health Monitoring: By utilizing geostatistical tools and methods, health professionals can map disease patterns and health outcomes to specific geographic locations (Sarkar & Sanyal, 2019). This spatial analysis helps in identifying hotspots of health issues prevalent among the Sundarbans tribals, such as vector-borne diseases like malaria and dengue (Hafiz Iqbal, 2020). Understanding these spatial distributions enables health authorities to allocate resources effectively and prioritize interventions, improving overall public health (Roy, 2019).

Enhanced Disaster Preparedness: The Sundarbans is highly susceptible to flooding and cyclones, events that significantly impact health and living conditions (Mozumdar et al., 2018). Integrating habitat and health perspectives using geostatistical insights allows for precise modeling of flood hazards and risk assessment (Dasgupta et al., 2021). This information is crucial for disaster preparedness, enabling communities to develop early warning systems, evacuation plans, and protective measures (Ahsan & Warner, 2014). Such proactive approaches can minimize injuries, fatalities, and health crises following natural disasters (Smit and Skinner, 2002).

Sustainable Resource Management: The relationship between health and environment is reciprocal (Gómez-Baggethun et al., 2012). Healthy ecosystems contribute to public health through the provision of clean water, air, and resources for food security (Carmin et al., 2012). Applying geostatistics to understand the dynamics of wetlands can enhance sustainable resource management, ensuring that tribal communities have continued access to vital natural resources (Almansi, Hardoy, & Pandiella, 2014). This is critical for their livelihoods and food security, which are closely tied to the availability of fish, shellfish, and other aquatic life (DiGregorio, 2013).

Advocacy and Policy Development: Geostatistical insights enable the visualization of health disparities and environmental challenges faced by the Sundarbans tribals (Sen, 2019). These powerful visual tools can be instrumental in advocacy efforts for policy changes at local, national, and international levels (Anwar et al., 2023). By illustrating the link between health and environmental conditions, stakeholders can push for protective legislation, conservation efforts, and investments in health infrastructure, fostering long-term improvements for tribal populations (Dey and Hazra, 2021).

Community Engagement and Empowerment: Involving tribal communities in data collection and interpretation can empower them and enhance their agency in decision-making processes (Jalais, 2007). By utilizing local knowledge alongside geostatistical methods, the community can better articulate their health needs, promote practices that support ecological balance, and engage in habitat conservation (Szabo et al., 2016). This participatory approach fosters a sense of ownership and responsibility toward both health and environment (Rahman et al., 2021).

CHALLENGES OF BRIDGING HEALTH AND HABITAT FOR SUNDARBANS TRIBALS

Despite the significant advantages of bridging health and habitat in the Sundarbans, several challenges hinder the effective integration of geostatistical insights into health policy and environmental management (Unogwu et al., 2024). Addressing these challenges is essential for fostering sustainability and protecting the rights to health for tribal communities (Dasgupta et al., 2014).

Data Scarcity and Quality: One of the primary challenges lies in the lack of quality data on health indicators and habitat dynamics (Danda & Sriskanthan, 2011). The Sundarbans region is often characterized by inadequate health facilities and limited access to reliable health information (Ghosh, 2012). The collection of high-quality geospatial and health data necessitates significant resources, which are often unavailable in rural and underfunded areas (Bhowmick and Poddar, 2023). This can lead to incomplete or biased analyses that might misrepresent the health landscape (Pitchaikani, 2020).

Cultural Sensitivity and Engagement: The Sundarbans tribals possess unique cultural beliefs and practices that influence their perceptions of health and the environment (StC,D. E., 2008). Engaging these communities without cultural insensitivity is paramount (Mahadevia & Vikas, 2012). Failure to do so can result in reluctance or resistance from community members to adopt health interventions or habitat management strategies (Afroz and Alam, 2013). Building trust and promoting inclusivity in health discussions is essential yet challenging in diverse societal contexts (Hiran et al., 2021).

Environmental Degradation and Climate Change: The Sundarbans face ongoing environmental challenges including deforestation, soil salinization, and the impacts of climate change, such as rising sea levels and intensified flooding (Doshi et al., 2024). These factors complicate efforts to integrate environmental health into community practices (Moore, 2023). As habitats change, traditional knowledge may no longer be applicable, necessitating continuous updates to health and habitat strategies (Mondal, Bowers, & Hossen, 2021).

Interdisciplinary Collaboration: Bridging health and habitat requires collaboration across multiple disciplines public health, environmental science, and social sciences which can be rife with institutional barriers (Islam, Dey, Sanyal, & Bandyopadhyay, 2022). Disparities in priorities, languages, and methodologies between fields can impede cohesive approaches to problem-solving (Ray, 2020). Fostering effective interdisciplinary collaboration is vital, yet often difficult to achieve (Begum, Lobry De Bruyn, Amirul Islam, & Kristiansen, 2022).

Limited Policy Implementation and Resources: Even when health and habitat strategies are developed using geostatistical insights, limited political will and financial resources can hinder implementation (Chanda & Akhand, 2023). Policymakers may prioritize short-term goals over the long-term health and environmental strategies needed for sustainability (Jamal, Ghosh, Hazarika, & Sen, 2022). Advocacy must therefore not only focus on raising awareness but also on securing funding and political support for necessary initiatives (Nicholls et al., 2020).

IMPLEMENTATIONS OF REGULATORY FRAMEWORKS

The Sundarbans, a unique mangrove ecosystem straddling India and Bangladesh, is home to diverse ethnic communities known as tribals (Mukhopadhyay, 2009). These communities often face significant health challenges exacerbated by flood hazards and dynamic wetland ecosystems (Biswas, 2013). The implementation of regulatory frameworks is essential for bridging health and habitat for these vulnerable populations (Thakur et al., 2021). Effective policy interventions can enhance health outcomes, build resilience to environmental changes, and promote sustainable livelihoods (Pramanik et al., 2021).

One crucial aspect of regulatory frameworks is integrating health and environmental policies (Sarwar Hossain et al., 2016). Regulatory bodies should promote collaboration among health, environment, and disaster management agencies to develop comprehensive strategies (Biswas & Nautiyal, 2020). The 2030 Agenda for Sustainable Development emphasizes the interlinkages between health and sustainable ecosystems (Adams et al., 2020). In the context of the Sundarbans, regulations can advocate for ecosystem-based approaches to public health wherein wetland conservation directly contributes to health improvements (Soltesova, Brown, Dayal, & Dodman, 2014). Such frameworks can include protecting crucial habitats from destructive practices like overfishing, unregulated tourism, and unsustainable agricultural methods that exacerbate flooding (Few et al., 2007).

Effective land-use regulations should be implemented to manage the balance between habitat conservation and human settlements (Roberts, 2008). By promoting zoning laws that limit construction in flood-prone areas, the risk of waterborne diseases and other health-related issues can be decreased (Roberts, 2013). This approach emphasizes the concept of "sustainable interactions," where the natural environment and human health do not compete but rather complement each other (Sivaramakrishnan, 2007). Such strategies could be supported through education and community engagement, equipping tribals with the knowledge to better understand and advocate for their rights and health in the context of their environment (Leck & Simon, 2012).

Moreover, financial instruments such as insurance schemes can be designed within regulatory frameworks to provide support to these communities during and after flood events (Kernaghan & da Silva, 2013). Implementing agricultural insurance can safeguard livelihoods and, consequently, health outcomes (Almansi, 2010). Financial backing for disaster relief can also ensure that immediate health needs are met in the aftermath of natural disasters (Almansi et al., 2014), reducing long-term impacts on the health of tribal populations. The involvement of indigenous knowledge in the formulation of these frameworks is crucial (Pelling & Manuel-Navarrete, 2011). Tribals have a deep understanding of their local ecosystems and can provide insights into sustainable practices that align with regulatory efforts (Honadle, 1999). Collaborative governance that includes tribal voices can enhance the acceptability and effectiveness of health and habitat policies (Adger, 2003).

Furthermore, continuous monitoring through geostatistical methods is vital for the evaluation of these regulations (Sahana et al., 2021). By utilizing geospatial data, authorities can identify hazards, track health outcomes, and create responsive strategies that evolve with changing environmental conditions (Uddin, 2019). It allows for a data-driven approach where interventions can be tailored to specific community needs (Okoye et al., 2024). Regulatory frameworks should also include health impact assessments for new policies or developments (Aggarwal et al., 2024). These assessments will help gauge the potential effects of environmental changes on community health and ensure that health considerations are integrated at every level not merely as an afterthought (Pal and Poddar, 2024).

CONCLUSION

The endeavor to bridge health and habitat for the Sundarbans tribals requires an intricate interplay of multiple disciplines, including public health, environmental science, and social policy (Tyagi et al., 2020). This unique ecosystem is not just a backdrop for human activity; it is intrinsically linked to the health, culture, and livelihoods of the tribal communities residing within it (Unnikrishnan and Shankar, 2007). The challenges posed by climate change, particularly through flood hazards and wetland dynamics, underscore the urgency for comprehensive strategies that prioritize the well-being of these communities (Polin & Alam, 2020).

The implementation of regulatory frameworks serves as a cornerstone for these strategies (Chanda & Akhand, 2023). As discussed, these frameworks ought to holistically integrate health considerations into environmental and disaster management policies (Francesco & Rahman, 2020). This integration is not merely a bureaucratic necessity but a moral obligation, considering the historical neglect that tribal communities have experienced (Ghosh, 2023). With environmental degradation, the risk of health crises increases, particularly from waterborne diseases and malnutrition (Mondal, Bowers, & Hossen, 2021). Hence, regulations must focus on sustainable practices that align with the intrinsic environmental knowledge of local tribes (Mitra, 2024).

Moreover, these regulatory frameworks should be adaptive (Alam & Bhardwaj, 2020). As geostatistical methods provide real-time data on flood risks and health indicators, regulatory bodies can better respond to emerging challenges, ensuring that interventions are dynamic and responsive to shifting conditions (Jamal, Ghosh, Hazarika, & Sen, 2022). This adaptability is essential not only for safeguarding health but also for fostering community resilience against future shocks (Begum, Lobry De Bruyn, Amirul Islam, & Kristiansen, 2022).

The role of community engagement cannot be understated (Mitra, 2024). Empowering tribals to participate in the creation, implementation, and evaluation of these regulatory frameworks ensures that the solutions are culturally appropriate and recognized as legitimate by the communities themselves (Mondal, Bowers, & Hossen, 2021). This collaborative governance approach strengthens the social fabric of tribal societies and engenders a sense of ownership over health outcomes (Francesco & Rahman, 2020).

Furthermore, investment in research and monitoring needs to be prioritized (Chanda & Akhand, 2023). Understanding the specific health impacts of environmental changes demands robust data, which can be achieved through partnerships with academic institutions and NGOs (Polin & Alam, 2020). These collaborations can augment the capacity for data collection and analysis, leading to more informed policies that truly address the needs of Sundarbans tribals (Tyagi et al., 2020).

SUGGESTIONS

The Sundarbans, a unique ecosystem characterized by its rich biodiversity and intricate waterways, is home to numerous tribal communities whose health and well-being are closely linked to their environmental conditions (Pal and Poddar, 2024). As climate change leads to increasing frequency and intensity of flood hazards, it is imperative to explore innovative and integrative solutions to uphold the right to health for Sundarbans tribals (Aggarwal et al., 2024).

Community-Centric Risk Assessment: Engaging local communities in geostatistical risk assessments can provide valuable insights into the specific vulnerabilities faced by Sundarbans tribals (Okoye et al., 2024). This participatory approach not only empowers the communities but also ensures that the identified risks are culturally relevant and context-specific (Uddin, 2019). Tools such as participatory mapping can be employed to document local knowledge on flood patterns, health issues, and traditional practices (Sahana et al., 2021).

Integrated Health and Environmental Surveillance: The creation of an integrated health and environmental monitoring system is crucial (Adams et al., 2020). This system should utilize geospatial technologies to track changes in flood dynamics and correlate them with health outcomes in tribal communities (Biswas & Nautiyal, 2020). Data collection can focus on waterborne diseases, mental health factors due to environmental stress, and nutrition based on access to wetlands (Pramanik et al., 2021).

Disaster Preparedness and Response Training: Building resilience through education and training programs on disaster preparedness is essential (Thakur et al., 2021). These programs should educate the tribes about environmental changes, emergency response strategies, and health care options during flood crises (Nicholls et al., 2020). Workshops can also foster knowledge sharing on traditional health practices that are adapted to changing environmental conditions (Jamal, Ghosh, Hazarika, & Sen, 2022).

Sustainable Livelihood Initiatives: Introducing sustainable livelihood programs that harness local resources while ensuring ecological balance can reduce vulnerability to flooding (Chanda & Akhand, 2023). Training programs on organic farming, fisheries management, and alternative income-generating activities can support health outcomes by improving nutrition and economic stability while conserving the wetlands (Begum, Lobry De Bruyn, Amirul Islam, & Kristiansen, 2022).

Policy Advocacy for Health Indices: Advocating for the inclusion of health and well-being indicators in environmental policies is necessary (Ray, 2020). Policymakers should consider health status as a significant factor in zoning and land-use planning, ensuring that tribal communities have the resources and protections needed to maintain their health amidst environmental stressors (Islam, Dey, Sanyal, & Bandyopadhyay, 2022).

Cultural Preservation Initiatives: Understanding the cultural significance of health and the environment to tribal groups can guide interventions (Mondal, Bowers, & Hossen, 2021). Preservation of traditional medicinal knowledge, coupled with modern healthcare access, can lead to more holistic healthcare models that respect and integrate tribal heritage (Moore, 2023).

Interdisciplinary Collaboration: Future projects should encourage collaboration among local governments, NGOs, public health experts, and environmental scientists to design multifaceted intervention programs (Hiran et al., 2021). This interdisciplinary approach ensures that health initiatives are grounded in environmental realities, further bridging the gap between health and habitat (Pitchaikani, 2020).

FUTURE RESEARCH DIRECTIONS

Moving forward, research targeting the Sundarbans and its tribal communities must be forward-thinking and tailored to address the complexities arising from climate change (Doshi et al., 2024).

Longitudinal Studies on Health Impacts of Flooding: Establishing long-term studies that explore the relationships between flood events, habitat degradation, and health outcomes will yield deeper insights into the cyclical nature of health challenges faced by Sundarbans tribals (Bhowmick and Poddar, 2023). Such studies can inform design improvements for future health interventions.

Integration of Traditional Ecological Knowledge (TEK) with Geostatistics: Future research should focus on integrating TEK with modern geostatistical methods to create comprehensive models of flood risks and health implications (Rahman et al., 2021). This fusion can bring beneficial insights into how tribal communities adapt their health practices according to environmental changes.

Impact Assessments of Climate Adaptation Strategies: Investigating the effectiveness of climate adaptation strategies at the community level will be essential. Researchers can evaluate how these strategies mitigate health risks associated with flooding and identify the gaps or challenges emanating from their implementation (Unogwu et al., 2024).

Advanced Modeling Techniques for Flood Prediction: Investing in research that develops advanced modeling techniques to better predict flooding events and their socio-health impacts can provide actionable insights (Dey and Hazra, 2021). Employing machine learning and AI along with traditional geostatistical methods could offer nuanced predictions based on historical data.

Ecosystem Services and Health Correlation Studies: Exploring the correlation between healthy wetlands defined by their ecosystem services and tribal health outcomes can emphasize the importance of wetland conservation (Anwar et al., 2023). This line of inquiry could lead to the development of policy frameworks that recognize the intrinsic value of ecosystems in maintaining public health.

Health Equity Research: Focus on health inequities faced by marginalized groups in the Sundarbans, assessing how environmental hazards exacerbate health disparities (Dasgupta et al., 2021). This research can inform targeted interventions that promote equity and access to health resources for tribal communities.

Resilience Building Frameworks: Developing frameworks that evaluate the resilience of Sundarbans tribal communities in the face of flood hazards can guide policy and practice (Satio et al., 2020). Such research can identify essential pillars of resilience, such as social networks, resource access, and adaptive capacities, which are crucial for safeguarding health in fluctuating environments.

REFERENCES

Adams, H., Adger, W. N., Ahmad, S., Ahmed, A., Begum, D., Matthews, Z., Rahman, M. M., Nilsen, K., Gurney, G. G., & Streatfield, P. K. (2020). Multi-dimensional well-being associated with economic dependence on ecosystem services in deltaic social-ecological systems of Bangladesh. *Regional Environmental Change*, *20*(42), 2–16. DOI: 10.1007/s10113-020-01620-x

Adger, W. N. (2003). *Social aspects of adaptive capacity*. Climate Change, Adaptive Capacity and Development, 29-49. DOI: 10.1142/9781860945816_0003

Adger, W. N., Dessai, S., Goulden, M., Hulme, M., Lorenzoni, I., Nelson, D. R., Naess, L. O., Wolf, J., & Wreford, A. (2009). Are there social limits to adaptation to climate change? *Climatic Change*, *93*(3-4), 335–354. DOI: 10.1007/s10584-008-9520-z

Adger, W. N., Hughes, T. P., Folke, C., Carpenter, S. R., & Rockström, J. (2009). Social-ecological resilience to coastal disasters. *Science*, *309*(5737), 1036–1039. DOI: 10.1126/science.1112122 PMID: 16099974

Aggarwal, K., Doshi, R., Mijwil, M. M., Hiran, K. K., Gök, M., & Bala, I. (2024). A Simple Way to Comprehend the Difference and the Significance of Artificial Intelligence in Agriculture. *Intelligent Techniques for Predictive Data Analytics*, 189-201.

Agrawal, A. (2008). *The role of local institutions in adaptation to climate change*. Social Dimensions of Climate Change: Equity and Vulnerability in a Warming World, 173-198. https://hdl.handle.net/10986/28274

Ahsan, M. N., & Warner, J. (2014). The socioeconomic vulnerability index: A pragmatic approach for assessing climate change led risks-A case study in the south-western coastal Bangladesh. *International Journal of Disaster Risk Reduction*, *8*, 32–49. DOI: 10.1016/j.ijdrr.2013.12.009

Al Mahmud, R., Titumir, M., Afrin, T., Islam, M. S., Titumir, R. A. M., Afrin, T., Islam, M. S., Onneshan, U., & Dhaka, B. (2020). *Traditional Knowledge, Institutions and Human Sociality in Sustainable Use and Conservation of Biodiversity of the Sundarbans of Bangladesh*. Managing Socio-ecological Production Landscapes and Seascapes for Sustainable Communities in Asia. Springer (pp. 67–91).

Alam, A. (2012). Some Indian bryophytes known for their biologically active compounds. *International Journal of Applied Biology and Pharmaceutical Technology*, *3*(2), 239–246.

Alam, M., & Bhardwaj, S. (2020). *Temporal variation in glacier's area and identification of glacial lakes in Sikkim*. DOI: 10.1007/978-981-15-2097-6_8

Ali, A. M. (2006). Rice to shrimp: Land use land cover changes and soil degradation in Southwestern Bangladesh. *Land Use Policy, 23*(4), 421–435. DOI: 10.1016/j.landusepol.2005.02.001

Ali, B. (2020). Community participation and quality of life in naturebased tourism: Exploring the antecedents and moderators. *Journal of Hospitality & Tourism Research (Washington, D.C.), 20*(10), 1–32. DOI: 10.1177/1096348020980094

Almansi, F. (2010). *The relationship between disaster risks and urban planning in Argentina,* Case study prepared for ISDR Global Assessment Report 2011.

Almansi, F., & Hardoy, J. (2013). *Climate change impact in coastal areas of the Rio de la Plata River: Actions for prevention and adaptation*. DOI: 10.4324/9781315658735

Almansi, F., Hardoy, J., & Pandiella, G. (2014). Impacto del cambio climático en ciudades costeras del estuario del Rio de la Plata. *Medio Ambiente y Urbanización,* [Impact of climate change on coastal cities in the estuary of the Rio de la Plata. Environment and Urbanization*], 79*, 217–237. DOI: 10.1080/17565529.2014.918868

Anguelovski, I., & Carmin, J. (2011). Something borrowed, everything new – current opinion: Innovation and institutionalisation in urban climate governance. *Current Opinion in Environmental Sustainability, 3*(3), 169–175. DOI: 10.1016/j.cosust.2010.12.017

Anurag Danda, A., Ghosh, N., Bandyopadhyay, J., & Hazra, S. (2019). Managed retreat: Adaptation to climate change in the Sundarbans ecoregion in the Bengal Delta. *Journal of the Indian Ocean Region, 15*(3), 317–335. DOI: 10.1080/19480881.2019.1652974

Anwar, G., Arianto, W., Simarmata, M., Bertham, Y. H., Yuwana, Y., Romeida, A., . . . Ruslan, D. (2023). Floristic Composition and Species Richness of Mangrove Ecosystems in Pantai Panjang–Pulau Baai Nature Park, Bengkulu Province, Indonesia. *International Journal of Advancement in Life Sciences Research*, 6(3), 10-18. 10-18. https://doi.org/DOI: DOI: 10.31632/ijalsr.2023.v06i03

Armitage, D. (2005). Adaptive capacity and community-based natural resource management. *Environmental Management, 35*(6), 703–715. DOI: 10.1007/s00267-004-0076-z PMID: 15940398

Ascoli, F. D. (1921). *A revenue history of the Sundarbans from 1870 to 1920*. Bengal Secretariat Book Depot, Calcutta. https://cslrepository.nvli.in//handle/123456789/7362

Ayeb-Karlsson, S., van der Geest, K., Ahmed, I., Huq, S., & Warner, K. (2016). A people-centred perspective on climate change, environmental stress, and livelihood resilience in Bangladesh. *Sustainability Science, 11*(4), 679–694. DOI: 10.1007/s11625-016-0379-z PMID: 30174739

Ayers, J. (2011). Resolving the adaptation paradox: Exploring the potential for deliberative adaption policy-making in Bangladesh. *Global Environmental Politics, 11*(1), 62–88. DOI: 10.1162/GLEP_a_00043

Ayers, J., & Forsyth, T. (2009). Community-based adaptation to climate change, strengthening resilience through development. *Environment, 51*(4), 23–31. DOI: 10.3200/ENV.51.4.22-31

Aziz, M. A., Tollington, S., Barlow, A., Goodrich, J., Shamsuddoha, M., Islam, M. A., & Groombridge, J. J. (2017). Investigating patterns of tiger and prey poaching in the Bangladesh Sundarbans: Implications for improved management. *Global Ecology and Conservation, 9*, 70–81. DOI: 10.1016/j.gecco.2016.12.001

Begum, F., Lobry De Bruyn, L., Amirul Islam, M., & Kristiansen, P. (2022). *The Role of Women's Participation in Co-Management of the Sundarban Mangrove Forest of Bangladesh*. University of New England. Retrieved from https://hdl.handle.net/1959.11/55652

Berger, R. M., & Kelly, J. J. (1993). Social Work in the Ecological Crisis. *Social Work, 38*, 521–526. https://www.jstor.org/stable/23717148

Berkes, F. (2009). Evolution of co-management: Role of knowledge generation, bridging organizations and social learning. *Journal of Environmental Management, 90*(5), 1692–1702. DOI: 10.1016/j.jenvman.2008.12.001 PMID: 19110363

Berkes, F. (2009). Indigenous ways of knowing and the study of environmental change. *Journal of the Royal Society of New Zealand, 39*(4), 151–156. DOI: 10.1080/03014220909510568

Berkes, F., & Ross, H. (2013). Community resilience: Toward an integrated approach. *Society & Natural Resources, 26*(1), 5–20. DOI: 10.1080/08941920.2012.736605

Besthorn, F. H., & McMillen, D. P. (2002). '*The Oppression of Women and Nature: Ecofeminism as a Framework for Social Justice Oriented Social Work*', Families and Society. *The Journal of Contemporary Human Service, 83*, 221–232. DOI: 10.1606/1044-3894.20

Beveridge, H. (1876). Favorite Share Flag Were the Sundarbans inhabited in ancient times. *Journal of the Asiatic Society of Bengal, 45*, •••. https://archive.org/details/biostor-281250

Bhowmik, D., & Poddar, S. (2023). Impacts of Flood in India from 1953 to 2020. [AMT]. *Advancement in Management and Technology*, *4*(2), 12–33.

Biesbroek, G. R., Klostermann, J. E. M., Termeer, C. J. A. M., & Kabat, P. (2013). On the nature of barriers to climate change adaptation. *Regional Environmental Change*, *13*(5), 1119–1129. DOI: 10.1007/s10113-013-0421-y

Biswas, B. (2013). "The God of Death Takes Half of Our Children": Health of Women and Children in the Sundarbans Islands. *Journal of Health Care for the Poor and Underserved*, 24(2), 730–740. https://doi.org/. 2013.0069DOI: 10.1353/hpu

Biswas, S., & Nautiyal, S. (2020). An assessment of socio-economic vulnerability at the household level: A study on villages of the Indian Sundarbans. *Environment, Development and Sustainability*, *23*(7), 11120–11137. DOI: 10.1007/s10668-020-01085-2

Breil, M., & Johnson, K. (2012). *Conceptualizing urban adaptation to climate change*. Review of Environment, Energy and Economics (Re3). Available at https://ssrn.com/abstract=2125728

Brown, A., Dayal, A., & Rumbaitis del Rio, C. (2012). From practice to theory: Emerging lessons from Asia for building urban climate change resilience. *Environment and Urbanization*, *24*(2), 531–556. DOI: 10.1177/0956247812456490

Bulkeley, H., & Tuts, R. (2013). Understanding urban vulnerability, adaptation and resilience in the context of climate change. *Local Environment*, *18*(6), 646–662. DOI: 10.1080/13549839.2013.788479

Cannon, T., & Mueller-Mahn, D. (2010). Vulnerability, resilience and development discourses in context of climate change. *Natural Hazards*, *55*(3), 621–635. DOI: 10.1007/s11069-010-9499-4

Carmin, J., Anguelovski, I., & Roberts, D. (2012). Urban climate adaptation in the global South: Planning in an emerging policy domain. *Journal of Planning Education and Research*, *32*(1), 18–32. DOI: 10.1177/0739456X11430951

Chakrabarti, R. (2009). *Local People and the Global Tiger: An Environmental History of the Sundarbans*. Global Environment, Vol. 2, (2009),(No.3), pp. 72-95. https://www.jstor.org/stable/43201488

Chanda, A., & Akhand, A. (2023, August). Challenges towards the Sustainability and Enhancement of the Indian Sundarban Mangrove's Blue Carbon Stock. *Life (Basel, Switzerland)*, *13*(8), 1–19. DOI: 10.3390/life13081787 PMID: 37629645

Chattopadhyay, H. (1999). *The Mystery of the Sunderbans* (First ed.). A Mukherjee & Co. https://archive.org/details/the-mystery-of-the-sundarbans/page/18/mode/2up

Chowdhury, A. N. (2019). Environment and wellbeing: Eco-psychiatry in Sundarban Delta, India. *International Journal of Management and Social Sciences*, *37*, 2. DOI: 10.21013/jmss.v14.n2sp.p5

Corfee-Morlot, J., Cochran, I., Hallegatte, S., & Teasdale, P. (2011). Multilevel risk governance and urban adaptation policy. *Climatic Change*, *104*(1), 169–197. DOI: 10.1007/s10584-010-9980-9

Cox, D., & Pawar, M. (2006). *International Social Work: Issues, Strategies and Programs*. SAGE., DOI: 10.4135/9781544308685

Cutter, S. L., Boruff, B. J., & Shirley, W. L. (2003). Social vulnerability to environmental hazards. *Social Science Quarterly*, *84*(2), 242–261. DOI: 10.1111/1540-6237.8402002

Danda, A. A. (2019). Environmental security in the Sundarban in the current climate change era: strengthening India-Bangladesh cooperation. ORF Occasional Paper, (220).

Das, S., & Hazra, S. (2020). Trapped or resettled: Coastal communities in the Sundarbans Delta. *Climate Crisis and Local Communities*, *64*, 15–17.

Dasgupta, S., Wheeler, D., & Ghosh, S. (2021). Fishing in Salty Waters: Poverty, Occupational Saline Exposure, and Women's Health in the Indian Sundarban. *Journal of Management and Sustainability*, *12*(1), 1. DOI: 10.5539/jms.v12n1p1

Dewulf, A. (2013). Contrasting frames in policy debates on climate change adaptation. *Wiley Interdisciplinary Reviews: Climate Change*, *4*(4), 321–330. DOI: 10.1002/wcc.227

Dey, M. K., & Hazra, A. K. (2021). Abundance of microarthropods population in different sites of Midnapore east coast of West Bengal, India. *International Journal of Advancement in Life Sciences Research*, 13-25. https://doi.org/DOI: DOI: 10.31632/ijalsr.2021.v04i03.003

DiGregorio, M. (2013a). Planning for climate resilience: Learning from grassroots, Presentation at *CBA 7 conference*, Dhaka, April 2013. DOI: 10.4324/9781315658735

Dodman, D., & Mitlin, D. (2013). Challenges for community-based adaptation: Discovering the potential for transformation. *Journal of International Development*, *25*(5), 640–659. DOI: 10.1002/jid.1772

Dodman, D., & Satterthwaite, D. (2008). Institutional capacity, climate change adaptation and the urban poor. *IDS Bulletin*, *39*(4), 67–74. DOI: 10.1111/j.1759-5436.2008.tb00478.x

Dominelli, L. (2010). *'Globalization, Contemporary Challenges and Social Work Practice'*. *International Social Work*, *53*(5), 599–612. DOI: 10.1177/0020872810371201

Dominelli, L. (2011). *'Climate Change: Social Workers' Roles and Contributions to Policy Debates and Interventions'*. *International Journal of Social Welfare*, *20*(4), 430–438. DOI: 10.1111/j.1468-2397.2011.00795.x

Doshi, R., Dadhich, M., Poddar, S., & Hiran, K. K. (2024). *Integrating generative AI in education to achieve sustainable development goals.*

Dubey, S., Trivedi, R., Chand, B., Mandal, B., & Rout, S. (2017). Farmers' perception of climate change, impacts on freshwater aquaculture and adaptation strategies in climatic change hotspots: A case of the Indian Sundarban delta. *Environmental Development*, *21*, 38–51. DOI: 10.1016/j.envdev.2016.12.002

Eakin, H., & Lemos, M. C. (2006). Adaptation and the state: Latin America and the challenge of capacity-building under globalization. *Global Environmental Change*, *16*(1), 7–18. DOI: 10.1016/j.gloenvcha.2005.10.004

Eakin, H., & Luers, A. L. (2006). Assessing the vulnerability of social-environmental systems. *Annual Review of Environment and Resources*, *31*(1), 365–394. DOI: 10.1146/annurev.energy.30.050504.144352

Elliot, D. (2012). *Social Development and Social Work.* In Healy, L. M., & Link, R. J. (Eds.), *Handbook of International Social Work* (pp. 102–108). Oxford University Press., DOI: 10.1093/acprof:oso/9780195333619.003.0014

Elliott, D. (1993). Social Work and Social Development: Towards an Integrative Model for Social Work Practice. *International Social Work*, *36*(1), 21–36. DOI: 10.1177/002087289303600103

Erickson, C. L. (2011). *Environmental Degradation and Preservation.* In Healy, L. M., & Link, R. J. (Eds.), *Handbook of International Social Work* (pp. 184–189). Oxford University Press., DOI: 10.1093/acprof:oso/9780195333619.003.0027

Eriksen, S. H., Brown, K., & Kelly, P. M. (2011). The dynamics of vulnerability: Locating coping strategies in Kenya and Tanzania. *The Geographical Journal*, *177*(4), 287–305. DOI: 10.1111/j.1475-4959.2005.00174.x

Eriksen, S. H., Nightingale, A. J., & Eakin, H. (2015). Reframing adaptation: The political nature of climate change adaptation. *Global Environmental Change*, *21*(2), 670–680. DOI: 10.1016/j.gloenvcha.2015.09.014

Few, R., Brown, K., & Tompkins, E. L. (2007). Public participation and climate change adaptation: Avoiding the illusion of inclusion. *Climate Policy*, *7*(1), 45–59. DOI: 10.1080/14693062.2007.9685637

Folke, C., Hahn, T., Olsson, P., & Norberg, J. (2005). Adaptive governance of social-ecological systems. *Annual Review of Environment and Resources*, *30*(1), 441–473. DOI: 10.1146/annurev.energy.30.050504.144511

Forsyth, T. (2013). Community-based adaptation: A review of past and future challenges. *Wiley Interdisciplinary Reviews: Climate Change*, *4*(5), 439–446. Advance online publication. DOI: 10.1002/wcc.231

Francesco, E. D., & Rahman, M. (2020, December). *Natural, social and economic regeneration in the Sundarbans of Bangladesh*. Circular Conversations. Retrieved from https://www.circularconversations.com/earth-heroes/bangladesh-environment-and-development-society

Germain, C. B. (1978). General-systems Theory and Ego Psychology: An Ecological Perspective. *The Social Service Review*, *52*(4), 535–550. https://www.jstor.org/stable/30015668. DOI: 10.1086/643676

Ghosh, A. (2012). *Living with changing climate–impact, vulnerability and adaptation challenges in Indian Sundarbans*. Centre for Science and Environment.

Ghosh, A. (2019). *The Tides of Life*. Indira Gandhi National Centre for the Arts.

Ghosh, O. (2023). Environmental hazards and disasters - A response towards mitigating disaster management. *AIP Conference Proceedings*, *2763*(1), 050008. DOI: 10.1063/5.0158892

Gómez-Baggethun, E., Corbera, E., & Reyes-García, V. (2012). Traditional ecological knowledge and global environmental change: Research findings and policy implications. *Ecology and Society*, *18*(4), 72. DOI: 10.5751/ES-06288-180472 PMID: 26097492

Government of West Bengal. (2020). (*rep.*). State Action Plan on Climate Change.

Gray, M. (2006). The Progress of Social Development in South Africa. *International Journal of Social Welfare*, *15*(s1, Suppl. 1), S53–S64. DOI: 10.1111/j.1468-2397.2006.00445.x

Gray, M., & Coates, J. R. (2012). Environmental Ethics for Social Work: Social Work's Responsibility to the Non-human World. *International Journal of Social Welfare*, *21*(3), 239–247. DOI: 10.1111/j.1468-2397.2011.00852.x

Gray, M., Mazibuko, F., & O'Brien, F. (1996). Social Work Education for Social Development. *Journal of Social Development in Africa*, *11*(1), 33–42. https://api.semanticscholar.org/CorpusID:221926574

Grothmann, T., & Patt, A. (2005). Adaptive capacity and human cognition: The process of individual adaptation to climate change. *Global Environmental Change*, *15*(3), 199–213. DOI: 10.1016/j.gloenvcha.2005.01.002

Gutierrez, K. S., & LePrevost, C. E. (2016). Climate justice in rural southeastern United States: A review of climate change impacts and effects on human health. *International Journal of Environmental Research and Public Health*, *13*(2), 189. Advance online publication. DOI: 10.3390/ijerph13020189 PMID: 26848673

Hafiz Iqbal, M. (2020). Valuing ecosystem services of Sundarbans mangrove forest for improved conservation: Approach of randomized conjoint experiment. *Forestry Economics Review*, *2*(1), 2631–3030. DOI: 10.1108/FER-04-2020-0008

Healy, L. M. (2008). *International Social Work: Professional Action in an Interdependent World* (2nd ed.). Oxford University Press., Available at https://global.oup.com/ushe/product/international-social-work-9780190922252?cc=in&lang=

Hiran, K. K., Khazanchi, D., Vyas, A. K., & Padmanaban, S. (Eds.). (2021). *Machine learning for sustainable development* (Vol. 9). Walter de Gruyter GmbH & Co KG., DOI: 10.1515/9783110702514

Honadle, G. (1999). *How Context Matters: Linking Environmental Policy to People and Place*. Kumarian Press., https://api.semanticscholar.org/CorpusID:152578133

Hurlimann, A. C., & March, A. P. (2012). The role of spatial planning in adapting to climate change. *Wiley Interdisciplinary Reviews: Climate Change*, *3*(5), 477–488. DOI: 10.1002/wcc.183

Islam, S. T., Dey, A., Sanyal, N., & Bandyopadhyay, S. (2022). *Troubles in The Sundarbans: A study of social and ecological issues in Hingalganj Block, North 24 Parganas West Bengal*. ActionAid International.

Jabeen, H., Johnson, C., & Allen, A. (2010). Built-in resilience: Learning from grassroots coping strategies for climate variability. *Environment and Urbanization*, *22*(2), 415–431. DOI: 10.1177/0956247810379937

Jahan, R. (1995). *The elusive agenda: Mainstreaming women in development*. London: ZED Books. https://www.jstor.org/stable/41260001

Jalais, A. (2007). The Sundarbans: Whose World Heritage Site? *Conservation & Society*, *5*(3), 335–342. https://www.jstor.org/stable/26392892

Jamal, S., Ghosh, A., Hazarika, R., & Sen, A. (2022, July). Livelihood, conflict and tourism: An assessment of livelihood impact in Sundarbans, West Bengal. *International Journal of Geoheritage and Parks*, *10*(1), 383–399. Advance online publication. DOI: 10.1016/j.ijgeop.2022.07.004

Kanjilal, B., Bose, S., Patra, N., Barman, D., Ghosh, U., Mandal, A., Vardrevu, L., & Sengupta, P. (2013). *How Healthy are the Children of Indian Sundarbans?*http://r4d.dfid.gov.uk/pdf/outputs/FutureHealth_RPC/SHWforweb.pdf

Kanjilal, B., Mazumdar, P., Mukherjee, M., Mondal, S., Barman, D., Singh, S., & Mandal, A. (2010). Health care in the Sundarbans (India): Challenges and plan for a better future. *Future Health Systems Research Programme* (Issues 1–91). DOI: 10.6133/apjcn.012016.01

Kanjilal, B., Mazumdar, P. G., Mukherjee, M., Mondal, S., Barman, D., & Singh, S. M. (2010, January). *Health care in the Sundarbans (India): Challenges. Future Health Systems Research Programme*, 21-30. Retrieved from https://assets.publishing.service.gov.uk/media/57a08b24ed915d3cfd000b68/sundarbans.pdf

Kemp, S. P., Whittaker, J. K., & Tracy, E. M. (1997) *Person–Environment Practice: The Social Ecology of Interpersonal Helping*. New York: Aldine De Gruyter. https://api.semanticscholar.org/CorpusID:141687227

Kernaghan, S., & da Silva, J. (2013). Initiating and sustaining action: Experiences building resilience to climate change in Asian cities. *Urban Climate*, *7*, 47–63. DOI: 10.1016/j.uclim.2013.10.008

Khan, S. A. R., Umar, M., Yu, Z., & Nawaz, M. T. (2023). A Recent Digitalization in Recycling Industry Attaining Ecological Sustainability: A Comprehensive Outlook and Future Trend. *Environmental Science and Pollution Research International*, *30*(47), 103760–103775. DOI: 10.1007/s11356-023-29537-y PMID: 37695483

Leck, H., & Simon, D. (2012). Fostering multiscalar collaboration and co-operation for effective governance of climate change adaptation. *Urban Studies (Edinburgh, Scotland)*, *50*(6), 1221–1238. Advance online publication. DOI: 10.1177/0042098012461675

Leichenko, R. (2011). Climate change and urban resilience. *Current Opinion in Environmental Sustainability*, *3*(3), 164–168. DOI: 10.1016/j.cosust.2010.12.014

Mahadevia, K., & Vikas, M. (2012). Climate change—Impact on the Sunderbans: A case study. *International Scientific Journal-Environmental Science*, *2*(1), 7–15.

Mertz, O., Mbow, C., Reenberg, A., & Diouf, A. (2009). Farmers' perceptions of climate change and agricultural adaptation strategies in rural Sahel. *Environmental Management*, *43*(5), 804–816. DOI: 10.1007/s00267-008-9197-0 PMID: 18810526

Midgley, J. (1994). Defining Social Development: Historical Trends and Conceptual Formulations. *Social Development Issues*, *16*(3), 3–19. DOI: 10.4135/9781446221839.n2

Midgley, J. (1995). *Social Development: The Developmental Perspective in Social Welfare*. SAGE., DOI: 10.4135/9781446221839.n2

Miller, S. E., Hayward, R. A., & Shaw, T. V. (2012). Environmental Shifts for Social Work: A Principles Approach. *International Journal of Social Welfare*, *21*(3), 270–277. DOI: 10.1111/j.1468-2397.2011.00848.x

Mimi, R., & Haynes, L. (2018). *The role of livelihood activities and income portfolios in shaping women's perceptions of wellbeing and adaptation choices, in the Indian Bengal Delta Statement of Originality.*

Mishra, M., Acharyya, T., Santos, C. A. G., da Silva, R. M., Kar, D., Mustafa Kamal, A. H., & Raulo, S. (2021). Geo-ecological impact assessment of severe cyclonic storm Amphan on Sundarban mangrove forest using geospatial technology. *Estuarine, Coastal and Shelf Science*, *260*, 107486. DOI: 10.1016/j.ecss.2021.107486

Mistri, A. (2019). Is the migration from Indian Sundarban an environmental migration? Investigating through Sustainable Livelihood Approach (SLA). *Asian Profile*, *47*(3), 195–218.

Mitlin, D., & Satterthwaite, D. (2012). Addressing poverty and inequality new forms of urban governance in Asia. *Environment and Urbanization*, *24*(2), 395–401. DOI: 10.1177/0956247812458062

Mitra, R. (2024). *Nature-based Solutions can Help Sundarbans Survive. DownToEarth*. Retrieved from https://www.downtoearth.org.in/news/climate-change/nature-based-solutions-can-help-sundarbans -survive-95230

Mizanur Rahman, M., Rakib Hossain, M., & Nazrul Islam, M. (2021). Climate Change Impact on Sundarbans: Challenges for Mitigation Strategies. *Climate Change in Bangladesh: A Cross-Disciplinary Framework*, 47-64.

Molinari, N. (2017). Intensifying Insecurities: The impact of climate change on vulnerability to human trafficking in the Indian Sundarbans. *Anti Trafficking Review, 363*(8), 1–11. DOI: 10.14197/atr.20121784

Mondal, M. K., Bhuiyan, S. I., & Franco, D. T. (2001). Soil salinity reduction and prediction of salt dynamics in the coastal rice lands of Bangladesh. *Agricultural Water Management, 47*(1), 9–23. DOI: 10.1016/S0378-3774(00)00098-6

Mondal, T., Bowers, W. W., & Hossen, M. S. (2021, March 25). Sustainable Management of Sundarbans: Stakeholder Attitudes Towards Participatory Management and Conservation of Mangrove Forests. *Journal of Sustainable Development, 14*(3), 23–40. DOI: 10.5539/jsd.v14n3p23

Moore, J. (2023, March 6). *PayBitoPro saves Sundarbans' child education.* Retrieved from PayBitoPro: https://www.paybito.com/lost-in -the -mangroves -paybito -steps -up -to -save -child -education -in -sundarbans/ 1

Moser, S. C., & Ekstrom, J. A. (2010). A framework to diagnose barriers to climate change adaptation. *Proceedings of the National Academy of Sciences of the United States of America, 107*(51), 22026–22031. DOI: 10.1073/pnas.1007887107 PMID: 21135232

Mozumdar, M., Shamsuzzaman, M., Un-Nabi, R., & Harun-Al-Rashid, A. (2018). Socio-economic characteristics and fishing operation activities of the artisanal fishers in the Sundarbans Mangrove Forest, Bangladesh. *Turkish Journal of Fisheries and Aquatic Sciences, 18*(6), 788–799. DOI: 10.4194/1303-2712-v18_6_05

Mukhopadhyay, A. (2009). *Cyclone Aila and the Sundarbans: An Enquiry into the Disaster and Politics of Aid and Relief.* In Research Report on CRG Annual Winter Course on Forced Migration.

Nicholls, R. J., Adger, W. N., Hutton, C. W., & Hanson, S. E. (2020). *Deltas in the Anthropocene.* Palgrave Macmillan Springer Nature., DOI: 10.1007/978-3-030-23517-8

Norgaard, K. M. (2017, September). The Sociological Imagination in a Time of Climate Change. *Global and Planetary Change, Elsevier.* DOI: DOI: 10.1016/j.gloplacha.2017.09.018

Norris, F., Stevens, S., Pfefferbaum, B., Wyche, K., & Pfefferbaum, R. (2008). Community resilience as metaphor, theory, set of capacities, and strategy for disaster readiness. *American Journal of Community Psychology, 41*(1–2), 127–150. DOI: 10.1007/s10464-007-9156-6 PMID: 18157631

Nyong, A., Adesina, F., & Elasha, B. O. (2007). The value of indigenous knowledge in climate change mitigation and adaptation strategies in the African Sahel. *Mitigation and Adaptation Strategies for Global Change*, *12*(5), 787–797. DOI: 10.1007/s11027-007-9099-0

O'Brien, K. (2012). Global environmental change II: From adaptation to deliberate transformation. *Progress in Human Geography*, *36*(5), 667–676. DOI: 10.1177/0309132511425767

O'Brien, K., Eriksen, S., Nygaard, L. P., & Schjolden, A. (2007). Why different interpretations of vulnerability matter in climate change discourses. *Climate Policy*, *7*(1), 73–88. DOI: 10.1080/14693062.2007.9685639

Okoye, K., Nganji, J. T., Hiran, K. K., & Hosseini, S. (2024, May). Impact and implications of AI methods and tools for the future of education. [). Frontiers Media SA.]. *Frontiers in Education*, *9*, 1434052. DOI: 10.3389/feduc.2024.1434052

Olhoff, A., & Schaer, C. (2010). *Screening tools and guidelines to support the mainstreaming of climate change adaptation into development assistance a stocktaking report.* New York: UNDP. http://www.undp.org/climatechange/library.shtml

Omer, S. (1979). *'Social Development'*. *International Social Work*, *22*(3), 11–26. DOI: 10.1177/002087287902200303

Pal, P., & Poddar, S. (2024). Applications of Smart Farming Sensors: A Way Forward. *Data Science for Agricultural Innovation and Productivity*, 175.

Panda, S., Sadhu, C., Pramanik, G., Pahari, S., & Hossain, J. (2016). Concerning public health situation of under-nutrition in children and anemia in women in Indian Sundarbans delta: A community based cross sectional investigation. *BMC Nutrition*, *2*(1), 1–10. DOI: 10.1186/s40795-016-0105-3

Peeters, J. (2012). The Place of Social Work in Sustainable Development: Towards Ecosocial Practice. *International Journal of Social Welfare*, *21*(3), 287–298. DOI: 10.1111/j.1468-2397.2011.00856.x

Pelling, M. (2003). *The vulnerability of cities: Natural disasters and social resilience*. Earthscan., DOI: 10.1080/15693430500111884

Pelling, M. (2011). *Adaptation to climate change: From resilience to transformation*. Routledge., DOI: 10.4324/9780203889046

Pelling, M., & Manuel-Navarrete, D. (2011). From resilience to transformation: The adaptive cycle in two Mexican urban centers. *Ecology and Society*, *16*(2), 11. DOI: 10.5751/ES-04038-160211

Pervin, M., Sultana, S., Phirum, A., Camara, I., Nzau, M., Phonnasane, V., & Anderson, S. (2013). *A framework for mainstreaming climate resilience into development planning*. IIED Working Paper, IIED, London https://www.iied.org/10050iied

Pitchaikani, J. S. (2020, March). Integrated coastal zone management practices for Sundarbans, India. *Indian Journal of Geo-Marine Sciences*, *49*(03), 352–356. https://core.ac.uk/download/pdf/326041666.pdf

Polin, F., & Alam, D. (2020, March 13). *Enhancing Community Participation in Forest Management: Case Study of the Sundarbans, Bangladesh*. South-West Region (Khulna Division). Retrieved from https://www.researchgate.net/publication/339916369_Enhancing_Community_Participation_in_Forest_Management_Case_Study_of_the_Sundarbans_Bangladesh

Pramanik, M., Szabo, S., Pal, I., Udmale, P., Pongsiri, M., & Chilton, S. (2021). Population health risks in multi-hazard environments: Action needed in the Cyclone Amphan and COVID-19 - hit Sundarbans region, India. *Climate and Development*, *14*(2), 99–104. DOI: 10.1080/17565529.2021.1889948

Rahman, M., Giedraitis, V. R., Lieberman, L. S., Akhtar, T., & Taminskienė, V. (2013). Shrimp cultivation with water salinity in Bangladesh: The implications of an ecological model. *Universal Journal of Public Health*, *1*(3), 131–142. DOI: 10.13189/ujph.2013.010313

Ray, P. S. (2020). Socio-Economic History in Sundarbans. *Journal of Critical Reviews*, *7*(2), 1030–1034.

Reid, H., Alam, M., Berger, R., Cannon, T., Huq, S., & Milligan, A. (2009). *Community-based adaptation to climate change: An overview*. Participatory Learning and Action: Community Based Adaptation to Climate Change, 60, 11–33. Available at https://www.researchgate.net/publication/272791313_Community-based_adaptation_to_climate_change_An_overview

Ribot, J. C., Najam, A., & Watson, G. (2009). *Climate Variation, Vulnerability and Sustainable Development in the Semi-arid Tropics*. In Schipper, E. L. F., & Burton, I. (Eds.), *The Earthscan Reader on Adaptation to Climate Change* (pp. 117–160). Earthscan., Available at https://www.researchgate.net/publication/291177311_Climate_variation_vulnerability_and_sustainable_development_in_the_semi-arid_tropics

Roberts, D. (2008). Thinking globally, acting locally institutionalising climate change at the the local government level in Durban, South Africa. *Environment and Urbanization*, *20*(2), 521–537. DOI: 10.1177/0956247808096126

Roberts, D. (2013). *Busting the Myths – An African reality check on Community Based Adaptation in Durban, South Africa*. Presentation at *CBA 7 conference*, Dhaka, April 2013. DOI: 10.1177/0956247808096126

Rogers, P. P., Jalal, K. F., & Boyd, J. A. (2007). *An Introduction to Sustainable Development*. Harvard University Press., DOI: 10.4324/9781849770477

Roy, D. (2020). 'On the Horns of a Dilemma'! Climate change, forest conservation and the marginal people in Indian Sundarbans. *Forum for Development Studies*, 47(2), 307–326. https://doi.org/. 1786452DOI: 10.1080/08039410.2020

Roy, M., Hulme, D., & Jahan, F. (2013). Contrasting adaptation responses by squatters and low-income tenants in Khulna, Bangladesh. *Environment and Urbanization*, *25*(1), 157–176. DOI: 10.1177/0956247813477362

Roy, S. (2018). Livelihood resilience of the indigenous Munda community in the Bangladesh Sundarbans Forest. *Handbook Of Climate Change Resilience*, 10(1), 1–21. DOI: 10.1007/978-3-319-71025-9_10-1

Roy, S. (2019). *Women's Use of the Sundarbans Forest Resources*. Springer Briefs in Environment, Security. *Development and Peace*, *29*, 59–71. DOI: 10.1007/978-981-13-6776-2_4

Rudra, A., & Chattopadhyay, A. (2018). *Environmental vulnerabilities and risks in population: A comparison of two Sundarban Delta*. Population Association of America, May, 1–14. https://www.researchgate.net/publication/326201083

Rudra, A., & Chattopadhyay, A. (2019). Environmental change of coastal Sundarbans: Impact on livelihood and standard of living status of indigenous people. *Environmental Quality Management*, *29*(2), 77–84. DOI: 10.1002/tqem.21667

Sahana, M., Rehman, S., Ahmed, R., & Sajjad, H. (2021). Analyzing climate variability and its effects in Sundarban Biosphere Reserve, India: Reaffirmation from local communities. *Environment, Development and Sustainability*, *23*(2), 2465–2492. DOI: 10.1007/s10668-020-00682-5

Saif, S., Russell, A. M., Nodie, S. I., Inskip, C., Lahann, P., Barlow, A., Greenwood Barlow, C., Islam, M. A., & MacMillan, D. C. (2016). Local usage of tiger parts and its role in tiger killing in the Bangladesh Sundarbans. *Human Dimensions of Wildlife*, *21*(2), 95–110. DOI: 10.1080/10871209.2015.1107786

Saito, O., Subramanian, S., Hashimoto, S., & Takeuchi, K. (2020). *Managing socio-ecological production landscapes and seascapes for sustainable communities in Asia mapping and navigating stakeholders, policy and action*. Springer https://www.springer.com/series/11884

Sarkar, S., & Sanyal, P. (2019). Community perception towards tourism development in the Indian Sundarbans: Case study of Jharkhali Island. *International Journal of Engineering Research & Technology (Ahmedabad)*, *8*(12), 526–540. www.ijert.org

Sarwar Hossain, M., Eigenbrod, F., Amoako Johnson, F., & Dearing, J. A. (2016). Unravelling the interrelationships between ecosystem services and human wellbeing in the Bangladesh delta. *International Journal of Sustainable Development and World Ecology*, *24*(2), 120–134. DOI: 10.1080/13504509.2016.1182087

Satterthwaite, D., Huq, S., Reid, H., Pelling, M., & Romero Lankao, P. (2007). *Adapting to climate change in urban areas: the possibilities and constraints in low- and middle-income nations*. Human Settlements Working Paper, IIED, London. https://www.iied.org/10549iied

Satterthwaite, D., & Mitlin, D. (2014). *Reducing urban poverty in the global south*. Routledge., DOI: 10.4324/9780203104316

Schipper, E. L., Ayers, J., Reid, H., Huq, S., & Rahman, A. (Eds.), *Community-based adaptation to climate change: scaling it up* (pp. 214–225). Routledge., DOI: 10.4324/9780203105061

Sen, H. S. (2019). *The Sundarbans: A disaster-prone ecoregion*. Berlin: Springer. 10, 978-3 https://link.springer.com/content/pdf/10.1007/978-3-030-00680-8.pdf

Sharma, D., & Tomar, S. (2010). Mainstreaming climate change adaptation in Indian cities. *Environment and Urbanization*, *22*(2), 451–465. DOI: 10.1177/0956247810377390

Sharma, R., Webster, P., & Bhattacharyya, S. (2014). Factors affecting the performance of community health workers in India: A multi-stakeholder perspective. *Global Health Action*, *7*(1), 25352. DOI: 10.3402/gha.v7.25352 PMID: 25319596

Sivaramakrishnan, K. C. (2007). *Democracy in urban India*. Urban Age. https://urbanage.lsecities.net/essays/democracy-in-urban-india

Smit, B., & Pilifosova, O. (2003). *From adaptation to adaptive capacity and vulnerability reduction.* In Climate Change, Adaptive Capacity and Development (pp. 9-28). Imperial College Press. DOI: 10.1142/9781860945816_0002

Smit, B., & Skinner, M. W. (2002). Adaptation options in agriculture to climate change: A typology. *Mitigation and Adaptation Strategies for Global Change*, *7*(1), 85–114. DOI: 10.1023/A:1015862228270

Smit, B., & Wandel, J. (2006). Adaptation, adaptive capacity and vulnerability. *Global Environmental Change*, *16*(3), 282–292. DOI: 10.1016/j.gloenvcha.2006.03.008

Soltesova, K., Brown, A., Dayal, A., & Dodman, D. (2014). *Community participation in urban adaptation to climate change: Potential and limits for CBA approaches*. Community-Based Adaptation to Climate Change., DOI: 10.4324/9780203105061

Sovacool, B. K. (2011). An international comparison of four polycentric approaches to climate and energy governance. *Energy Policy, 39*(6), 3832–3844. DOI: 10.1016/j.enpol.2011.04.014

Sovacool, B. K. (2018). Bamboo beating bandits: Conflict, inequality, and vulnerability in the political ecology of climate change adaptation in Bangladesh. *World Development, 102*, 183–194. DOI: 10.1016/j.worlddev.2017.10.014

StC. (2008). D. E. *Convention on the Conservation of Migratory Species of Wild Animals*.

Sundarbans National park. (1987). *UNESCO World Heritage Centre*. https://whc.unesco.org/en/list/452

Syam, D. (2013). *National engagement and replication in Indonesia*. Presentation at the CBA7 conference, Dhaka, April 2013. DOI: 10.1177/0956247808096126

Szabo, S., Hossain, M. S., Adger, W. N., Matthews, Z., Ahmed, S., Lázár, A. N., & Ahmad, S. (2016). Soil salinity, household wealth and food insecurity in tropical deltas: Evidence from south-westcoast of Bangladesh. *Sustainability Science, 11*(3), 411–421. DOI: 10.1007/s11625-015-0337-1 PMID: 30174734

Thakur, S., Maity, D., Mondal, I., Basumatary, G., Ghosh, P. B., Das, P., & De, T. K. (2021). Assessment of changes in land use, land cover, and land surface temperature in the mangrove forest of Sundarbans, northeast coast of India. *Environment, Development and Sustainability, 23*(2), 1917–1943. DOI: 10.1007/s10668-020-00656-7

Tripodi, T., & Potocky-Tripodi, M. (2007). *International Social Work Research: Issues and Prospects*. Oxford University Press., DOI: 10.1093/acprof:oso/9780195187250.001.0001

Twigg, J. (2007). *Characteristics of a disaster-resilient community, a guidance note*. Available at https://www.researchgate.net/publication/315799865_Characteristics_of_a_Disaster-resilient_Community_A_Guidance_Note_Characteristics_of_a_Disaster-resilient_Community_A_Guidance_Note

Tyagi, S. K. S., Mukherjee, A., Pokhrel, S. R., & Hiran, K. K. (2020). An intelligent and optimal resource allocation approach in sensor networks for smart agri-IoT. *IEEE Sensors Journal, 21*(16), 17439–17446. DOI: 10.1109/JSEN.2020.3020889

Tyler, S., & Moench, M. (2012). A framework for urban climate resilience. *Climate and Development*, *4*(4), 311–326. DOI: 10.1080/17565529.2012.745389

Uddin, S. M. (2019). Religion, Nature, and Life in the Sundarbans. *Asian Ethnology*, *78*(2), 289–310. https://www.jstor.org/stable/26845329

Unogwu, O. J., Doshi, R., Hiran, K. K., Mijwil, M. M., Catherine, A. T., & Abotaleb, M. (2024). *Exploring the Implications of Emerging Artificial Intelligence Technologies at Edge Computing in Higher Education*. In Integrating Generative AI in Education to Achieve Sustainable Development Goals (pp. 169-182). IGI Global. DOI: DOI: 10.4018/979-8-3693-2440-0.ch009

Yohe, G., & Tol, R. S. J. (2002). Indicators for social and economic coping capacity moving toward a working definition of adaptive capacity. *Global Environmental Change*, *12*(1), 25–40. DOI: 10.1016/S0959-3780(01)00026-7

ADDITIONAL READING

Andtfolk, M., Nyholm, L., Eide, H., Rauhala, A., & Fagerström, L. (2022). Attitudes Toward the Use of Humanoid Robots in Healthcare—A Cross-Sectional Study. *AI & Society*, *37*(4), 1739–1748. DOI: 10.1007/s00146-021-01271-4

Asiedu-Berkoe, F., Chandi, G. M., Bandoh, D. A., Atsu, B. K., Lokossou, V. K., Antara, S. N., Sarkodie, B., & Kenu, E. (2022). State of Public Health Emergency Preparedness and Response Capacity of Ghana. *Journal of Interventional Epidemiology and Public Health*, *5*(4), 1–16. DOI: 10.37432/jieph.supp.2022.5.4.05.4

Chauhan, P., Bali, A., & Kaur, S. (2024). Breaking Barriers for Accessible Health Programs: The Role of Telemedicine in a Global Healthcare Transformation. In *Transformative Approaches to Patient Literacy and Healthcare Innovation*. IGI Global., DOI: 10.4018/979-8-3693-3661-8.ch014

KEY TERMS AND DEFINITIONS

Climate: Adaptation: Climate adaptation refers to strategies and measures implemented to help communities and ecosystems adjust to the adverse effects of climate change.

Climate Justice: Environmental justice is a social movement and legal principle that advocates for fair treatment and meaningful involvement of all individuals, particularly marginalized communities, in environmental decision-making.

Health Equity: Health equity refers to the principle that everyone should have a fair opportunity to attain their highest level of health, regardless of social or economic status.

Indigenous Rights: Indigenous rights refer to the collective and individual entitlements of Indigenous peoples, recognizing their unique cultural identities, traditional lands, and ways of life.

Traditional Ecological Knowledge (TEK): Traditional Ecological Knowledge encompasses the understanding and practices developed by indigenous communities over generations concerning their local ecosystems. TEK includes insights about biodiversity, resource management, and sustainable practices that can inform contemporary environmental approaches and policies, contributing to both conservation and sustainable development.

Compilation of References

Abbas, M., Akhai, S., Abbas, U., Jafri, R., & Arif, S. M. (2025). Ai-enabled sustainable urban planning and management. In *Real-World Applications of AI Innovation* (pp. 233–260). IGI Global Scientific Publishing.

Adams, H., Adger, W. N., Ahmad, S., Ahmed, A., Begum, D., Matthews, Z., Rahman, M. M., Nilsen, K., Gurney, G. G., & Streatfield, P. K. (2020). Multi-dimensional well-being associated with economic dependence on ecosystem services in deltaic social-ecological systems of Bangladesh. *Regional Environmental Change*, *20*(42), 2–16. DOI: 10.1007/s10113-020-01620-x

Adelekan, I. O. (2020). Urban dynamics, everyday hazards and disaster risks in Ibadan, Nigeria. *Environment and Urbanization*, *32*(1), 213–232. Advance online publication. DOI: 10.1177/0956247819844738

Adger, W. N. (2003). *Social aspects of adaptive capacity*. Climate Change, Adaptive Capacity and Development, 29-49. DOI: 10.1142/9781860945816_0003

Adger, W. N., Dessai, S., Goulden, M., Hulme, M., Lorenzoni, I., Nelson, D. R., Naess, L. O., Wolf, J., & Wreford, A. (2009). Are there social limits to adaptation to climate change? *Climatic Change*, *93*(3-4), 335–354. DOI: 10.1007/s10584-008-9520-z

Adger, W. N., Hughes, T. P., Folke, C., Carpenter, S. R., & Rockström, J. (2009). Social-ecological resilience to coastal disasters. *Science*, *309*(5737), 1036–1039. DOI: 10.1126/science.1112122 PMID: 16099974

Adiono, T., Muttaqin, R., Rahayu, R. M., & Putra, R. V. W. (2020). *Cloud System Design using AMQP Protocol for Smart Devices System Applications*. DOI: 10.31227/osf.io/z824j

Aggarwal, K., Doshi, R., Mijwil, M. M., Hiran, K. K., Gök, M., & Bala, I. (2024). A Simple Way to Comprehend the Difference and the Significance of Artificial Intelligence in Agriculture. *Intelligent Techniques for Predictive Data Analytics*, 189-201.

Agrawal, A. (2008). *The role of local institutions in adaptation to climate change.* Social Dimensions of Climate Change: Equity and Vulnerability in a Warming World, 173-198. https://hdl.handle.net/10986/28274

Ahsan, M. N., & Warner, J. (2014). The socioeconomic vulnerability index: A pragmatic approach for assessing climate change led risks-A case study in the south-western coastal Bangladesh. *International Journal of Disaster Risk Reduction*, *8*, 32–49. DOI: 10.1016/j.ijdrr.2013.12.009

Ajzen, I. (1991). The theory of planned behavior. *Organizational Behavior and Human Decision Processes*, *50*(2), 179–211. DOI: 10.1016/0749-5978(91)90020-T

Ajzen, I. (2002). Perceived behavioral control, self-efficacy, locus of control, and the theory of planned behavior 1. *Journal of Applied Social Psychology*, *32*(4), 665–683. DOI: 10.1111/j.1559-1816.2002.tb00236.x

Ajzen, I., & Madden, T. J. (1986). Prediction of goal-directed behavior: Attitudes, intentions, and perceived behavioral control. *Journal of Experimental Social Psychology*, *22*(5), 453–474. DOI: 10.1016/0022-1031(86)90045-4

Akinsemolu, A. A., Onyeaka, H. N., & Tamasiga, P. (2024). Climate-smart agriculture as a possible solution to mitigate climate change impact on food security in Sub-Saharan Africa. In *Food and Energy Security* (Vol. 13, Issue 1). DOI: 10.1002/fes3.509

Al Mahmud, R., Titumir, M., Afrin, T., Islam, M. S., Titumir, R. A. M., Afrin, T., Islam, M. S., Onneshan, U., & Dhaka, B. (2020). *Traditional Knowledge, Institutions and Human Sociality in Sustainable Use and Conservation of Biodiversity of the Sundarbans of Bangladesh.* Managing Socio-ecological Production Landscapes and Seascapes for Sustainable Communities in Asia. Springer (pp. 67–91).

Alahakoon, D., Nawaratne, R., Xu, Y., De Silva, D., Sivarajah, U., & Gupta, B. (2023). Self-Building Artificial Intelligence and Machine Learning to Empower Big Data Analytics in Smart Cities. *Information Systems Frontiers*, *25*(1), 221–240. DOI: 10.1007/s10796-020-10056-x

Alam, M., & Bhardwaj, S. (2020). *Temporal variation in glacier's area and identification of glacial lakes in Sikkim.* DOI: 10.1007/978-981-15-2097-6_8

Alam, A. (2012). Some Indian bryophytes known for their biologically active compounds. *International Journal of Applied Biology and Pharmaceutical Technology*, *3*(2), 239–246.

Alexander, D. E. (2020). *Natural disasters*. Routledge., DOI: 10.4324/9781003071372

Alghamdi, M. (2024). Smart city urban planning using an evolutionary deep learning model. *Soft Computing*, *28*(1), 447–459. DOI: 10.1007/s00500-023-08219-4

Ali, B., & Awad, A. I. (2018). Cyber and Physical Security Vulnerability Assessment for IoT-Based Smart Homes. *Sensors 2018, Vol. 18, Page 817, 18*(3), 817. DOI: 10.3390/s18030817

Ali, A. M. (2006). Rice to shrimp: Land use land cover changes and soil degradation in Southwestern Bangladesh. *Land Use Policy*, *23*(4), 421–435. DOI: 10.1016/j.landusepol.2005.02.001

Ali, B. (2020). Community participation and quality of life in naturebased tourism: Exploring the antecedents and moderators. *Journal of Hospitality & Tourism Research (Washington, D.C.)*, *20*(10), 1–32. DOI: 10.1177/1096348020980094

Almalki, F. A., Alsamhi, S. H., Sahal, R., Hassan, J., Hawbani, A., Rajput, N. S., Saif, A., Morgan, J., & Breslin, J. (2021). Green IoT for Eco-Friendly and Sustainable Smart Cities: Future Directions and Opportunities. *Mobile Networks and Applications 2021 28:1, 28*(1), 178–202. DOI: 10.1007/s11036-021-01790-w

Almansi, F. (2010). *The relationship between disaster risks and urban planning in Argentina,* Case study prepared for ISDR Global Assessment Report 2011.

Almansi, F., & Hardoy, J. (2013). *Climate change impact in coastal areas of the Rio de la Plata River: Actions for prevention and adaptation*. DOI: 10.4324/9781315658735

Almansi, F., Hardoy, J., & Pandiella, G. (2014). Impacto del cambio climático en ciudades costeras del estuario del Rio de la Plata. *Medio Ambiente y Urbanización*, [Impact of climate change on coastal cities in the estuary of the Rio de la Plata. Environment and Urbanization*]*, *79*, 217–237. DOI: 10.1080/17565529.2014.918868

Almolhis, N., Alashjaee, A. M., Duraibi, S., Alqahtani, F., & Moussa, A. N. (2020). The Security Issues in IoT-Cloud: A Review. *Proceedings - 2020 16th IEEE International Colloquium on Signal Processing and Its Applications, CSPA 2020*, 191–196. DOI: 10.1109/CSPA48992.2020.9068693

Almuhanna, A. (2018). Smart Traffic Management System for Traffic Control using Automated Mechanical and Electronic Devices. *IOP Conference Series. Materials Science and Engineering*, *377*(1), 012201. DOI: 10.1088/1757-899X/377/1/012201

Alston, J. M., & Venner, R. J. (2002). The effects of the US Plant Variety Protection Act on wheat genetic improvement. *Research Policy, 31*(4), 527–542. DOI: 10.1016/S0048-7333(01)00123-8

Altieri, M. A. (1995). *Agroecology: The science of sustainable agriculture*. Westview Press. DOI: 10.3362/9781788532310

Altieri, M. A. (2009). Agroecology, small farms, and food sovereignty. *Monthly Review (New York, N.Y.), 61*(3), 102–113. DOI: 10.14452/MR-061-03-2009-07_8

Anami, S., De Block, M., MacHuka, J., & Van Lijsebettens, M. (2009). Molecular improvement of tropical maize for drought stress tolerance in Sub-Saharan Africa. *Critical Reviews in Plant Sciences, 28*(1–2), 16–35. Advance online publication. DOI: 10.1080/07352680802665305

Andersen, R. (2008). *Governing Agrobiodiversity: Plant Genetics and Developing Countries*. Ashgate.

Andersen, R., & Winge, T. (2012). *Realising Farmers' Rights to Crop Genetic Resources: Success Stories and Best Practices*. Routledge.

Anguelovski, I., & Carmin, J. (2011). Something borrowed, everything new – current opinion: Innovation and institutionalisation in urban climate governance. *Current Opinion in Environmental Sustainability, 3*(3), 169–175. DOI: 10.1016/j.cosust.2010.12.017

Anurag Danda, A., Ghosh, N., Bandyopadhyay, J., & Hazra, S. (2019). Managed retreat: Adaptation to climate change in the Sundarbans ecoregion in the Bengal Delta. *Journal of the Indian Ocean Region, 15*(3), 317–335. DOI: 10.1080/19480881.2019.1652974

Anwar, G., Arianto, W., Simarmata, M., Bertham, Y. H., Yuwana, Y., Romeida, A., . . . Ruslan, D. (2023). Floristic Composition and Species Richness of Mangrove Ecosystems in Pantai Panjang–Pulau Baai Nature Park, Bengkulu Province, Indonesia. *International Journal of Advancement in Life Sciences Research*, 6(3), 10-18. 10-18. https://doi.org/DOI: DOI: 10.31632/ijalsr.2023.v06i03

Apte, J. S., Messier, K. P., Gani, S., Brauer, M., Kirchstetter, T. W., Lunden, M. M., & Marshall, J. D. (2017). High-resolution air pollution mapping with Google Street View cars: Exploiting big data. *Environmental Science & Technology, 51*(12), 6999–7008. DOI: 10.1021/acs.est.7b00891 PMID: 28578585

Apthorpe, N., Reisman, D., Sundaresan, S., Narayanan, A., & Feamster, N. (2017). *Spying on the Smart Home: Privacy Attacks and Defenses on Encrypted IoT Traffic*. https://arxiv.org/abs/1708.05044v1

Armitage, D. (2005). Adaptive capacity and community-based natural resource management. *Environmental Management*, *35*(6), 703–715. DOI: 10.1007/s00267-004-0076-z PMID: 15940398

Ascoli, F. D. (1921). *A revenue history of the Sundarbans from 1870 to 1920*. Bengal Secretariat Book Depot, Calcutta. https://cslrepository.nvli.in//handle/123456789/7362

Asif, M., Ersoy, A., Lodhi, R. N., & Cobanoglu, C. (2025). Mapping the metaverse-sustainability nexus in hospitality and tourism: A bibliometric approach. *Online Information Review*, *49*(3), 600–622. DOI: 10.1108/OIR-09-2023-0510

Askari, H., Khajepour, A., Khamesee, M. B., & Wang, Z. L. (2019). Embedded self-powered sensing systems for smart vehicles and intelligent transportation. *Nano Energy*, *66*, 104103. DOI: 10.1016/j.nanoen.2019.104103

Atta, A., Abbas, S., Khan, M. A., Ahmed, G., & Farooq, U. (2020). An adaptive approach: Smart traffic congestion control system. *Journal of King Saud University. Computer and Information Sciences*, *32*(9), 1012–1019. DOI: 10.1016/j.jksuci.2018.10.011

Ayeb-Karlsson, S., van der Geest, K., Ahmed, I., Huq, S., & Warner, K. (2016). A people-centred perspective on climate change, environmental stress, and livelihood resilience in Bangladesh. *Sustainability Science*, *11*(4), 679–694. DOI: 10.1007/s11625-016-0379-z PMID: 30174739

Ayers, J. (2011). Resolving the adaptation paradox: Exploring the potential for deliberative adaption policy-making in Bangladesh. *Global Environmental Politics*, *11*(1), 62–88. DOI: 10.1162/GLEP_a_00043

Ayers, J., & Forsyth, T. (2009). Community-based adaptation to climate change, strengthening resilience through development. *Environment*, *51*(4), 23–31. DOI: 10.3200/ENV.51.4.22-31

Aziz, M. A., Tollington, S., Barlow, A., Goodrich, J., Shamsuddoha, M., Islam, M. A., & Groombridge, J. J. (2017). Investigating patterns of tiger and prey poaching in the Bangladesh Sundarbans: Implications for improved management. *Global Ecology and Conservation*, *9*, 70–81. DOI: 10.1016/j.gecco.2016.12.001

Bagley, M. A. (2020). Digital sequence information: Towards a legal framework to ensure fair and equitable benefit sharing. *International Journal of Law and Technology*, *28*(3), 259–283. DOI: 10.1093/ijlit/eaaa011

Baker, J., Ebrahim, Z., & Zarina, O. (2019). *Data challenges for climate resilience: Empowering action through better information*. World Bank Group.

Baldwin, A. (2009). Legal ecology: Ecosystem function and the reconfiguration of property. *Environmental Politics*, *18*(4), 588–606.

Banđur, Đ., Jakšić, B., Banđur, M., & Jović, S. (2019). An analysis of energy efficiency in Wireless Sensor Networks (WSNs) applied in smart agriculture. *Computers and Electronics in Agriculture*, *156*, 500–507. DOI: 10.1016/j.compag.2018.12.016

Barocas, S., Hardt, M., & Narayanan, A. (2019). *Fairness and machine learning: Limitations and opportunities.* https://fairmlbook.org

Batty, M. (2018). Digital twins. *Environment and Planning. B, Urban Analytics and City Science*, *45*(5), 817–820. DOI: 10.1177/2399808318796416

Beck, M. B., Balbi, S., & Lanzanova, D. (2020). Bridging the gap between data and decision-making: Leveraging artificial intelligence in support of sustainable development. *Environmental Modelling & Software*, *132*, 104808. DOI: 10.1016/j.envsoft.2020.104808

Begum, F., Lobry De Bruyn, L., Amirul Islam, M., & Kristiansen, P. (2022). *The Role of Women's Participation in Co-Management of the Sundarban Mangrove Forest of Bangladesh*. University of New England. Retrieved from https://hdl.handle.net/1959.11/55652

Bemmaor, A. C. (1995). Predicting behavior from intention-to-buy measures: The parametric case. *JMR, Journal of Marketing Research*, *32*(2), 176–191. DOI: 10.1177/002224379503200205

Benegal, V., Rajkumar, R. P., & Muralidharan, K. (2008). Does areca nut use lead to dependence? *Drug and Alcohol Dependence*, *97*(1-2), 114–121. DOI: 10.1016/j.drugalcdep.2008.03.016 PMID: 18490113

Berger, R. M., & Kelly, J. J. (1993). Social Work in the Ecological Crisis. *Social Work*, *38*, 521–526. https://www.jstor.org/stable/23717148

Berkes, F. (2009). Evolution of co-management: Role of knowledge generation, bridging organizations and social learning. *Journal of Environmental Management*, *90*(5), 1692–1702. DOI: 10.1016/j.jenvman.2008.12.001 PMID: 19110363

Berkes, F. (2009). Indigenous ways of knowing and the study of environmental change. *Journal of the Royal Society of New Zealand*, *39*(4), 151–156. DOI: 10.1080/03014220909510568

Berkes, F., & Ross, H. (2013). Community resilience: Toward an integrated approach. *Society & Natural Resources*, *26*(1), 5–20. DOI: 10.1080/08941920.2012.736605

Besthorn, F. H., & McMillen, D. P. (2002). '*The Oppression of Women and Nature: Ecofeminism as a Framework for Social Justice Oriented Social Work*', Families and Society. *The Journal of Contemporary Human Service*, *83*, 221–232. DOI: 10.1606/1044-3894.20

Beveridge, H. (1876). Favorite Share Flag Were the Sundarbans inhabited in ancient times. *Journal of the Asiatic Society of Bengal*, *45*, •••. https://archive.org/details/biostor-281250

Bhatia, M., & Ahanger, T. A. (2021). Intelligent decision-making in Smart Food Industry: Quality perspective. *Pervasive and Mobile Computing*, *72*, 101304. Advance online publication. DOI: 10.1016/j.pmcj.2020.101304

Bhowmik, D., & Poddar, S. (2023). Impacts of Flood in India from 1953 to 2020. [AMT]. *Advancement in Management and Technology*, *4*(2), 12–33.

Biesbroek, G. R., Klostermann, J. E. M., Termeer, C. J. A. M., & Kabat, P. (2013). On the nature of barriers to climate change adaptation. *Regional Environmental Change*, *13*(5), 1119–1129. DOI: 10.1007/s10113-013-0421-y

Biswas, B. (2013). "The God of Death Takes Half of Our Children": Health of Women and Children in the Sundarbans Islands. *Journal of Health Care for the Poor and Underserved*, 24(2), 730–740. https://doi.org/. 2013.0069DOI: 10.1353/hpu

Biswas, S., & Nautiyal, S. (2020). An assessment of socio-economic vulnerability at the household level: A study on villages of the Indian Sundarbans. *Environment, Development and Sustainability*, *23*(7), 11120–11137. DOI: 10.1007/s10668-020-01085-2

Boer, B. (2016). *Environmental Law Dimensions of Human Rights*. Oxford University Press.

Boopathi, S. (2024). Advancements in Machine Learning and AI for Intelligent Systems in Drone Applications for Smart City Developments. Https://Services.Igi-Global.Com/Resolvedoi/Resolve.Aspx?Doi=10.4018/978-1-6684-9596-4.Ch002, 15–45. DOI: 10.4018/978-1-6684-9596-4.ch002

Borowiak, C. (2004). Farmers' rights: Intellectual property regimes and the struggle over seeds. *Politics & Society*, *32*(4), 511–543. DOI: 10.1177/0032329204269979

Bosnjak, M., Ajzen, I., & Schmidt, P. (2020). The theory of planned behavior: Selected recent advances and applications. *Europe's Journal of Psychology*, *16*(3), 352–356. DOI: 10.5964/ejop.v16i3.3107 PMID: 33680187

Boudreau, M. E., DeVries, B., & Wessels, K. (2021). Transfer learning for environmental hazard mapping: Challenges and opportunities. *Remote Sensing*, *13*(12), 2345. DOI: 10.3390/rs13122345

Breil, M., & Johnson, K. (2012). *Conceptualizing urban adaptation to climate change*. Review of Environment, Energy and Economics (Re3). Available at https://ssrn.com/abstract=2125728

Brown, A., Dayal, A., & Rumbaitis del Rio, C. (2012). From practice to theory: Emerging lessons from Asia for building urban climate change resilience. *Environment and Urbanization*, *24*(2), 531–556. DOI: 10.1177/0956247812456490

Brush, S. B. (2005). *Farmers' Bounty: Locating Crop Diversity in the Contemporary World*. Yale University Press.

Bukari, C., & Aluko, O. A. (2023). Severity of climate change and deprivation outcomes: Micro-level assessment for sub-Saharan Africa. *Environmental Science & Policy*, *150*, 103593. Advance online publication. DOI: 10.1016/j.envsci.2023.103593

Bulkeley, H., & Tuts, R. (2013). Understanding urban vulnerability, adaptation and resilience in the context of climate change. *Local Environment*, *18*(6), 646–662. DOI: 10.1080/13549839.2013.788479

Burrell, J. (2016). How the machine 'thinks': Understanding opacity in machine learning algorithms. *Big Data & Society*, *3*(1), 2053951715622512. DOI: 10.1177/2053951715622512

Butt, H. A., Ahad, A., Wasim, M., Shayea, I., Coelho, P. J., Pires, I. M., & Garcia, N. M. (2023). Federated Machine Learning In 5G Smart Healthcare: A Security Perspective Review. *Procedia Computer Science*, *224*, 580–586. DOI: 10.1016/j.procs.2023.09.084

Calo, R. (2016). Robotics and the lessons of cyberlaw. *California Law Review*, *103*(3), 513–563. DOI: 10.15779/Z38XX96

Campi, M. (2017). The effect of intellectual property rights on agricultural productivity. *Agricultural Economics*, *48*(3), 327–339. DOI: 10.1111/agec.12337

Campi, M., & Nuvolari, A. (2021). Intellectual property rights and agricultural development: Evidence from a worldwide index of IPRs in agriculture (1961-2018). *The Journal of Development Studies*, *57*(4), 650–668. DOI: 10.1080/00220388.2020.1817395

Cannon, T., & Mueller-Mahn, D. (2010). Vulnerability, resilience and development discourses in context of climate change. *Natural Hazards*, *55*(3), 621–635. DOI: 10.1007/s11069-010-9499-4

Capra, F., & Mattei, U. (2015). *The ecology of law: Toward a legal system in tune with nature and community*. Berrett-Koehler Publishers.

Carmin, J., Anguelovski, I., & Roberts, D. (2012). Urban climate adaptation in the global South: Planning in an emerging policy domain. *Journal of Planning Education and Research, 32*(1), 18–32. DOI: 10.1177/0739456X11430951

CBD. (1992). *Convention on Biological Diversity*. United Nations. https://www.cbd.int

Çelik, K. T., & Şekeroğlu, A. (2023). The importance of geographical information systems in urban and landscape planning: A bibliometric analysis. *Megaron, 18*(4).

Chakrabarti, R. (2009). *Local People and the Global Tiger: An Environmental History of the Sundarbans*. Global Environment, Vol. 2, (2009),(No.3), pp. 72-95. https://www.jstor.org/stable/43201488

Chalon, V. & Spoon, P. (2007). Trends in multi-document analysis.

Chanda, A., & Akhand, A. (2023, August). Challenges towards the Sustainability and Enhancement of the Indian Sundarban Mangrove's Blue Carbon Stock. *Life (Basel, Switzerland), 13*(8), 1–19. DOI: 10.3390/life13081787 PMID: 37629645

Chang, V., Sharma, S., & Li, C. S. (2020). Smart cities in the 21st century. *Technological Forecasting and Social Change, 153*, 119447. DOI: 10.1016/j.techfore.2018.09.002

Chattopadhyay, H. (1999). *The Mystery of the Sunderbans* (First ed.). A Mukherjee & Co. https://archive.org/details/the-mystery-of-the-sundarbans/page/18/mode/2up

Chaudhary, B. (2013). *Protecting Farmers' Rights in the Global IPR Regime: Challenges and Strategies*. Springer.

Chaudhary, S., Gustafson, D., Mathys, A., & Altieri, M. (2019). A systems approach to assessing agroecology's role in sustainable food systems. *Nature Sustainability, 2*(5), 393–400.

Cheng, G., Han, J., & Lu, X. (2017). Remote Sensing Image Scene Classification by Deep Learning. *IEEE Transactions on Geoscience and Remote Sensing, 55*(5), 2670–2681.

Cheng, X., Hu, Y., & Yu, G. (2022). AI in wildfire prediction and monitoring: A review. *International Journal of Wildland Fire, 31*(7), 617–627. DOI: 10.1071/WF21035

Chen, J. (2005). The parable of the seeds: Interpreting the Plant Variety Protection Act in furtherance of innovation policy. *The Notre Dame Law Review*, *81*, 105.

Chen, J., Ramanathan, L., & Alazab, M. (2021). Holistic big data integrated artificial intelligent modeling to improve privacy and security in data management of smart cities. *Microprocessors and Microsystems*, *81*, 103722. DOI: 10.1016/j.micpro.2020.103722

Chen, Y. Y., Lin, Y. H., Kung, C. C., Chung, M. H., & Yen, I. H. (2019). Design and implementation of cloud analytics-assisted smart power meters considering advanced artificial intelligence as edge analytics in demand-side management for smart homes. *Sensors (Basel)*, *19*(9), 2047. DOI: 10.3390/s19092047 PMID: 31052502

Chen, Y., Wang, Y., & Zhou, D. (2021). Knowledge map of urban morphology and thermal comfort: A bibliometric analysis based on citespace. *Buildings (Basel, Switzerland)*, *11*(10), 427. DOI: 10.3390/buildings11100427

Chhetri, N., Easterling, W., Terando, A., & Udry, C. (2022). Intellectual property rights and the global governance of agricultural biotechnology. *World Development*, *151*, 105757. DOI: 10.1016/j.worlddev.2021.105757

Chiarolla, C., & Louafi, S. (2015). Implementing the Multilateral System of Access and Benefit-Sharing: Lessons from the Crop Case. *Biodiversity and Conservation*, *24*(13), 3119–3142.

Chilunjika, A., & Gumede, N. (2021). Climate Change and Human Security in Sub-Saharan Africa. *African Renaissance, 2021*(si1). DOI: 10.31920/2516-5305/v2021sin1a2

Chollet, F. (2021). *Deep learning with Python* (2nd ed.). Manning Publications.

Chowdhury, A. N. (2019). Environment and wellbeing: Eco-psychiatry in Sundarban Delta, India. *International Journal of Management and Social Sciences*, *37*, 2. DOI: 10.21013/jmss.v14.n2sp.p5

Cicea, C. (2020). Mapping the research agenda of energy in urban areas: A bibliometric analysis. *Theoretical and Empirical Researches in Urban Management*, *15*(2), 66–80.

Clancy, M. S., & Moschini, G. (2017). Intellectual property rights and the ascent of proprietary innovation in agriculture. *Annual Review of Resource Economics*, *9*(1), 53–74. DOI: 10.1146/annurev-resource-100516-053524

Cleveland, D. A., & Soleri, D. (2007). *Farmers, Scientists, and Plant Breeding: Integrating Knowledge and Practice*. CABI.

Codjoe, S. N. A., & Atiglo, D. Y. (2020). The Implications of Extreme Weather Events for Attaining the Sustainable Development Goals in Sub-Saharan Africa. In *Frontiers in Climate* (Vol. 2). DOI: 10.3389/fclim.2020.592658

Cohen, E. & Foku, A. (1995). Classification of beautiful handwritten text IAP02.

Conner, M., & Armitage, C. J. (1998). Extending the theory of planned behavior: A review and avenues for further research. *Journal of Applied Social Psychology*, *28*(15), 1429–1464. DOI: 10.1111/j.1559-1816.1998.tb01685.x

Coombe, R. J. (2009). The recognition of indigenous peoples' and community traditional knowledge in international law. *St. Thomas Law Review*, *21*(2), 275–298.

Corfee-Morlot, J., Cochran, I., Hallegatte, S., & Teasdale, P. (2011). Multilevel risk governance and urban adaptation policy. *Climatic Change*, *104*(1), 169–197. DOI: 10.1007/s10584-010-9980-9

Correa, C. M. (2000). *Intellectual Property Rights, the WTO and Developing Countries: The TRIPS Agreement and Policy Options*. Zed Books.

Correa, C. M. (2013). *Plant Variety Protection in Developing Countries: A Tool for Designing a Sui Generis Plant Variety Protection System*. GRAIN.

Cox, D., & Pawar, M. (2006). *International Social Work: Issues, Strategies and Programs*. SAGE., DOI: 10.4135/9781544308685

Croucher, R., & Islam, S. (2002). Socio-economic aspects of areca nut use. *Addiction Biology*, *7*(1), 139–146. DOI: 10.1080/13556210120091518 PMID: 11900634

Cui, X. (2018). How can cities support sustainability: A bibliometric analysis of urban metabolism. *Ecological Indicators*, *93*, 704–717. DOI: 10.1016/j.ecolind.2018.05.056

Cullet, P. (2005). Environmental justice in the use and conservation of biodiversity and water: The role of local communities. *Journal of Environmental Law*, *17*(2), 185–210.

Cullet, P., & Mayet, M. (2011). The sui generis system of seed rights in India: Its implementation and potential. *Journal of Law. Environmental Development*, *7*(2), 248–271.

Curtis, J. M. (2012). Intellectual property rights and international trade: An overview.

Cuthbert, M. O., Taylor, R. G., Favreau, G., Todd, M. C., Shamsudduha, M., Villholth, K. G., MacDonald, A. M., Scanlon, B. R., Kotchoni, D. O. V., Vouillamoz, J. M., Lawson, F. M. A., Adjomayi, P. A., Kashaigili, J., Seddon, D., Sorensen, J. P. R., Ebrahim, G. Y., Owor, M., Nyenje, P. M., Nazoumou, Y., & Kukuric, N. (2019). Observed controls on resilience of groundwater to climate variability in sub-Saharan Africa. *Nature*, *572*(7768), 230–234. Advance online publication. DOI: 10.1038/s41586-019-1441-7 PMID: 31391559

Cutter, S. L. (2006). The geography of social vulnerability: Race, class, and catastrophe. *Social Science Quarterly*, *87*(2), 286–291. DOI: 10.1111/j.1540-6237.2006.00384.x

Cutter, S. L., Boruff, B. J., & Shirley, W. L. (2003). Social vulnerability to environmental hazards. *Social Science Quarterly*, *84*(2), 242–261. DOI: 10.1111/1540-6237.8402002

Cutter, S. L., Burton, C. G., & Emrich, C. T. (2008). Disaster resilience indicators for benchmarking baseline conditions. *Journal of Homeland Security and Emergency Management*, *5*(1). Advance online publication. DOI: 10.2202/1547-7355.1732

D'Alpaos, C., & Andreolli, F. (2020). Urban quality in the city of the future: A bibliometric multicriteria assessment model. *Ecological Indicators*, *117*, 106575. DOI: 10.1016/j.ecolind.2020.106575

Danda, A. A. (2019). Environmental security in the Sundarban in the current climate change era: strengthening India-Bangladesh cooperation. ORF Occasional Paper, (220).

Dasgupta, S., Wheeler, D., & Ghosh, S. (2021). Fishing in Salty Waters: Poverty, Occupational Saline Exposure, and Women's Health in the Indian Sundarban. *Journal of Management and Sustainability*, *12*(1), 1. DOI: 10.5539/jms.v12n1p1

Das, S., & Hazra, S. (2020). Trapped or resettled: Coastal communities in the Sundarbans Delta. *Climate Crisis and Local Communities*, *64*, 15–17.

De Jonge, B. (2014). *Plant variety protection and farmers' rights: Towards a new balance*. Wageningen Academic Publishers.

De Perez, E. C., Stephens, E., Bischiniotis, K., Van Aalst, M., Van Den Hurk, B., Mason, S., Nissan, H., & Pappenberger, F. (2017). Should seasonal rainfall forecasts be used for flood preparedness? *Hydrology and Earth System Sciences*, *21*(9), 4517–4524. Advance online publication. DOI: 10.5194/hess-21-4517-2017

De Schutter, O. (2009). *Seed policies and the right to food: Enhancing agrobiodiversity and encouraging innovation*. Report of the Special Rapporteur on the right to food, UN Doc. A/64/170.

De Vries, H., Dijkstra, M., & Kuhlman, P. (1988). Self-efficacy: The third factor besides attitude and subjective norm as a predictor of behavioural intentions. *Health Education Research*, *3*(3), 273–282. DOI: 10.1093/her/3.3.273

Delmer, D. P., Nottenburg, C., Graff, G. D., & Bennett, A. B. (2003). Intellectual property resources for international development in agriculture. *Plant Physiology*, *133*(4), 1666–1670. DOI: 10.1104/pp.103.028993 PMID: 14681529

Dewulf, A. (2013). Contrasting frames in policy debates on climate change adaptation. *Wiley Interdisciplinary Reviews: Climate Change*, *4*(4), 321–330. DOI: 10.1002/wcc.227

Dey, M. K., & Hazra, A. K. (2021). Abundance of microarthropods population in different sites of Midnapore east coast of West Bengal, India. *International Journal of Advancement in Life Sciences Research*, 13-25. https://doi.org/DOI: DOI: 10.31632/ijalsr.2021.v04i03.003

Di Febbraro, A., Gallo, F., Giglio, D., & Sacco, N. (2020). Traffic management system for smart road networks reserved for self-driving cars. *IET Intelligent Transport Systems*, *14*(9), 1013–1024. DOI: 10.1049/iet-its.2019.0675

Djègbè, I., Zinsou, M., Dovonou, E. F., Tchigossou, G., Soglo, M., Adéoti, R., Gbaguidi, B., Atoyebi, S., Chandre, F., Akogbéto, M., Lines, J., & Djouaka, R. (2020). Minimal tillage and intermittent flooding farming systems show a potential reduction in the proliferation of Anopheles mosquito larvae in a rice field in Malanville, Northern Benin. *Malaria Journal*, *19*(1), 333. Advance online publication. DOI: 10.1186/s12936-020-03406-2 PMID: 32928223

Dodman, D., & Mitlin, D. (2013). Challenges for community-based adaptation: Discovering the potential for transformation. *Journal of International Development*, *25*(5), 640–659. DOI: 10.1002/jid.1772

Dodman, D., & Satterthwaite, D. (2008). Institutional capacity, climate change adaptation and the urban poor. *IDS Bulletin*, *39*(4), 67–74. DOI: 10.1111/j.1759-5436.2008.tb00478.x

Dominelli, L. (2010). '*Globalization, Contemporary Challenges and Social Work Practice*'. *International Social Work*, *53*(5), 599–612. DOI: 10.1177/0020872810371201

Dominelli, L. (2011). *'Climate Change: Social Workers' Roles and Contributions to Policy Debates and Interventions'*. *International Journal of Social Welfare*, *20*(4), 430–438. DOI: 10.1111/j.1468-2397.2011.00795.x

Doshi, R., Dadhich, M., Poddar, S., & Hiran, K. K. (2024). *Integrating generative AI in education to achieve sustainable development goals.*

Dossou-Yovo, E. R., Devkota, K. P., Akpoti, K., Danvi, A., Duku, C., & Zwart, S. J. (2022). Thirty years of water management research for rice in sub-Saharan Africa: Achievement and perspectives. In *Field Crops Research* (Vol. 283). DOI: 10.1016/j.fcr.2022.108548

Dossou-Yovo, E. R., Kouadio, S. A. K., & Saito, K. (2023). Effects of mid-season drainage on iron toxicity, rice yield, and water productivity in irrigated systems in the derived savannah agroecological zone of West Africa. *Field Crops Research*, *296*, 108901. Advance online publication. DOI: 10.1016/j.fcr.2023.108901

Dubey, S., Trivedi, R., Chand, B., Mandal, B., & Rout, S. (2017). Farmers' perception of climate change, impacts on freshwater aquaculture and adaptation strategies in climatic change hotspots: A case of the Indian Sundarban delta. *Environmental Development*, *21*, 38–51. DOI: 10.1016/j.envdev.2016.12.002

Dutfield, G. (2008). *Turning plant varieties into intellectual property: The UPOV Convention and its implications for farmers*. Quaker United Nations Office.

Dutfield, G. (2011). *Food, Biological Diversity and Intellectual Property: The Role of the International Union for the Protection of New Varieties of Plants (UPOV)*. Quaker United Nations Office.

Eakin, H., & Lemos, M. C. (2006). Adaptation and the state: Latin America and the challenge of capacity-building under globalization. *Global Environmental Change*, *16*(1), 7–18. DOI: 10.1016/j.gloenvcha.2005.10.004

Eakin, H., & Luers, A. L. (2006). Assessing the vulnerability of social-environmental systems. *Annual Review of Environment and Resources*, *31*(1), 365–394. DOI: 10.1146/annurev.energy.30.050504.144352

Echendu, A. J. (2021). Women, Development, and Flooding Disaster Research in Nigeria: A Scoping Review. *Environmental and Earth Sciences Research Journal*, *8*(4), 147–152. Advance online publication. DOI: 10.18280/eesrj.080401

Elish, M. C. (2019). Moral crumple zones: Cautionary tales in human-robot interaction. *Engaging Science, Technology, and Society*, *5*, 40–60. DOI: 10.17351/ests2019.260

Elliot, D. (2012). *Social Development and Social Work*. In Healy, L. M., & Link, R. J. (Eds.), *Handbook of International Social Work* (pp. 102–108). Oxford University Press., DOI: 10.1093/acprof:oso/9780195333619.003.0014

Elliott, D. (1993). Social Work and Social Development: Towards an Integrative Model for Social Work Practice. *International Social Work, 36*(1), 21–36. DOI: 10.1177/002087289303600103

Eriksen, S. H., Brown, K., & Kelly, P. M. (2011). The dynamics of vulnerability: Locating coping strategies in Kenya and Tanzania. *The Geographical Journal, 177*(4), 287–305. DOI: 10.1111/j.1475-4959.2005.00174.x

Eriksen, S. H., Nightingale, A. J., & Eakin, H. (2015). Reframing adaptation: The political nature of climate change adaptation. *Global Environmental Change, 21*(2), 670–680. DOI: 10.1016/j.gloenvcha.2015.09.014

Esfandi, S., Tayebi, S., Byrne, J., Taminiau, J., Giyahchi, G., & Alavi, S. A. (2024). Smart Cities and Urban Energy Planning: An Advanced Review of Promises and Challenges. *Smart Cities 2024, Vol. 7, Pages 414-444, 7*(1), 414–444. DOI: 10.3390/smartcities7010016

Estrada, E., & Vargas, M. P. M. (2021). Data Science Process for Smart Cities. In Latin American Women and Research Contributions to the IT Field (pp. 348-370). IGI Global Scientific Publishing.

ETC Group. (2013). *Putting the Cartel before the Horse… and Farm, Seeds, Soil, Peasants, etc.*

FAO. (2009). *International Treaty on Plant Genetic Resources for Food and Agriculture*. Food and Agriculture Organization of the United Nations.

FAO. (2010). *The Second Report on the State of the World's Plant Genetic Resources for Food and Agriculture*.

FAO. (2015). Status of World's Soil resources. https://openknowledge.fao.org/server/api/core/bitstreams/6ec24d75-19bd-4f1f-b1c5-5becf50d0871/content

FAO. (2019). *The state of the world's biodiversity for food and agriculture*. FAO Commission on Genetic Resources for Food and Agriculture.

FAO. (2021). *Agroecology and biodiversity: A pathway to sustainable food systems*. FAO.

Farmers Rights. (2022, December). Why farmers rights matter? https://www.farmersrights.org/what-are-farmers-rights/why-farmers-rights-matter/

Feizi, A., Joo, S., Kwigizile, V., & Oh, J. S. (2020). A pervasive framework toward sustainability and smart-growth: Assessing multifaceted transportation performance measures for smart cities. *Journal of Transport & Health*, *19*, 100956. DOI: 10.1016/j.jth.2020.100956

Few, R., Brown, K., & Tompkins, E. L. (2007). Public participation and climate change adaptation: Avoiding the illusion of inclusion. *Climate Policy*, *7*(1), 45–59. DOI: 10.1080/14693062.2007.9685637

Filho, G. P. R., Meneguette, R. I., Maia, G., Pessin, G., Gonçalves, V. P., Weigang, L., Ueyama, J., & Villas, L. A. (2020). A fog-enabled smart home solution for decision-making using smart objects. *Future Generation Computer Systems*, *103*, 18–27. DOI: 10.1016/j.future.2019.09.045

Fjeld, J., Achten, N., Hilligoss, H., Nagy, A., & Srikumar, M. (2020). Principled artificial intelligence: Mapping consensus in ethical and rights-based approaches to principles for AI. *Berkman Klein Center Research Publication No. 2020-1*. DOI: 10.2139/ssrn.3518482

Floridi, L., Cowls, J., Beltrametti, M., Chatila, R., Chazerand, P., Dignum, V., Luetge, C., Madelin, R., Pagallo, U., Rossi, F., Schafer, B., Valcke, P., & Vayena, E. (2018). AI4People—An ethical framework for a good AI society: Opportunities, risks, principles, and recommendations. *Minds and Machines*, *28*(4), 689–707. DOI: 10.1007/s11023-018-9482-5 PMID: 30930541

Folke, C., Carpenter, S. R., Walker, B., Scheffer, M., Chapin, T., & Rockström, J. (2010). Resilience thinking: Integrating resilience, adaptability and transformability. *Ecology and Society*, *15*(4), 20. DOI: 10.5751/ES-03610-150420

Folke, C., Hahn, T., Olsson, P., & Norberg, J. (2005). Adaptive governance of social-ecological systems. *Annual Review of Environment and Resources*, *30*(1), 441–473. DOI: 10.1146/annurev.energy.30.050504.144511

Forsyth, T. (2013). Community-based adaptation: A review of past and future challenges. *Wiley Interdisciplinary Reviews: Climate Change*, *4*(5), 439–446. Advance online publication. DOI: 10.1002/wcc.231

Francesco, E. D., & Rahman, M. (2020, December). *Natural, social and economic regeneration in the Sundarbans of Bangladesh*. Circular Conversations. Retrieved from https://www.circularconversations.com/earth-heroes/bangladesh-environment-and-development-society

Frison, C., López, F., & Esquinas-Alcázar, J. (Eds.). (2011). *Plant genetic resources and food security: Stakeholder perspectives on the International Treaty on Plant Genetic Resources for Food and Agriculture*. Earthscan.

Fuglie, K., & MacDonald, J. M. (2023). Expanded intellectual property protections for crop seeds increase innovation and market power for companies. *Amber Waves: The Economics of Food, Farming, Natural Resources, and Rural America, 2023.*

Fu, X. (2024). Natural Language Processing in Urban Planning: A Research Agenda. *Journal of Planning Literature*, *39*(3), 395–407. DOI: 10.1177/08854122241229571

Fu, Y., & Zhang, X. (2017). Trajectory of urban sustainability concepts: A 35-year bibliometric analysis. *Cities (London, England)*, *60*, 113–123. DOI: 10.1016/j.cities.2016.08.003

Gannon, K. E., Conway, D., Pardoe, J., Ndiyoi, M., Batisani, N., Odada, E., Olago, D., Opere, A., Kgosietsile, S., Nyambe, M., Omukuti, J., & Siderius, C. (2018). Business experience of floods and drought-related water and electricity supply disruption in three cities in sub-Saharan Africa during the 2015/2016 El Niño. *Global Sustainability*, *1*, e14. Advance online publication. DOI: 10.1017/sus.2018.14

Gao, L., Zhang, L., & Chen, X. (2022). Deep learning in disaster emergency response: Applications and challenges. *Natural Hazards*, *113*(2), 1375–1400. DOI: 10.1007/s11069-022-05124-w

Geris, J., Comte, J. C., Franchi, F., Petros, A. K., Tirivarombo, S., Selepeng, A. T., & Villholth, K. G. (2022). Surface water-groundwater interactions and local land use control water quality impacts of extreme rainfall and flooding in a vulnerable semi-arid region of Sub-Saharan Africa. *Journal of Hydrology (Amsterdam)*, *609*, 127834. Advance online publication. DOI: 10.1016/j.jhydrol.2022.127834

Germain, C. B. (1978). General-systems Theory and Ego Psychology: An Ecological Perspective. *The Social Service Review*, *52*(4), 535–550. https://www.jstor.org/stable/30015668. DOI: 10.1086/643676

Geva, A., & Goldman, A. (1991). Satisfaction measurement in guided tours. *Annals of Tourism Research*, *18*(2), 177–185. DOI: 10.1016/0160-7383(91)90002-S

Ghijsen, H. C. (2002). Property rights on plant varieties: An overview. *Journal of New Seeds*, *4*(1-2), 195–212. DOI: 10.1300/J153v04n01_15

Ghosh, A. (2012). *Living with changing climate–impact, vulnerability and adaptation challenges in Indian Sundarbans*. Centre for Science and Environment.

Ghosh, A. (2019). *The Tides of Life*. Indira Gandhi National Centre for the Arts.

Ghosh, O. (2023). Environmental hazards and disasters - A response towards mitigating disaster management. *AIP Conference Proceedings*, *2763*(1), 050008. DOI: 10.1063/5.0158892

Gibson, J. (2004). *Community resources: Intellectual property, international trade and the protection of traditional knowledge*. Ashgate.

Girsberger, M. (1999). *Biodiversity and the Concept of Farmers' Rights in International Law*. Peter Lang.

Gkontzis, A. F., Kotsiantis, S., Feretzakis, G., & Verykios, V. S. (2024). Enhancing Urban Resilience: Smart City Data Analyses, Forecasts, and Digital Twin Techniques at the Neighborhood Level. *Future Internet 2024, Vol. 16, Page 47, 16*(2), 47. DOI: 10.3390/fi16020047

Godin, G., Valois, P., & Lepage, L. (1993). The pattern of influence of perceived behavioral control upon exercising behavior: An application of Ajzen's theory of planned behavior. *Journal of Behavioral Medicine*, *16*(1), 81–102. DOI: 10.1007/BF00844756 PMID: 8433359

Golubchikov, O., & Thornbush, M. (2020). Artificial Intelligence and Robotics in Smart City Strategies and Planned Smart Development. *Smart Cities 2020, Vol. 3, Pages 1133-1144, 3*(4), 1133–1144. DOI: 10.3390/smartcities3040056

Gómez-Baggethun, E., Corbera, E., & Reyes-García, V. (2012). Traditional ecological knowledge and global environmental change: Research findings and policy implications. *Ecology and Society*, *18*(4), 72. DOI: 10.5751/ES-06288-180472 PMID: 26097492

Gonzalez, R., & Woods, R. (2008). *Digital Image Processing*. Pearson Prentice Hall.

Goodfellow, I., Bengio, Y., & Courville, A. (2016). *Deep learning*. MIT Press.

Goodfellow, I., Bengio, Y., & Courville, A. (2016). *Deep Learning*. MIT Press.

Government of West Bengal. (2020). (*rep.*). State Action Plan on Climate Change.

GRAIN. (2005). *TRIPS-plus through the back door*.

Gray, M. (2006). The Progress of Social Development in South Africa. *International Journal of Social Welfare*, *15*(s1, Suppl. 1), S53–S64. DOI: 10.1111/j.1468-2397.2006.00445.x

Gray, M., & Coates, J. R. (2012). Environmental Ethics for Social Work: Social Work's Responsibility to the Non-human World. *International Journal of Social Welfare*, *21*(3), 239–247. DOI: 10.1111/j.1468-2397.2011.00852.x

Gray, M., Mazibuko, F., & O'Brien, F. (1996). Social Work Education for Social Development. *Journal of Social Development in Africa, 11*(1), 33–42. https://api.semanticscholar.org/CorpusID:221926574

Griffith, D. C., Kelly-Hope, L. A., & Miller, M. A. (2006). Review of reported cholera outbreaks worldwide, 1995-2005. In *American Journal of Tropical Medicine and Hygiene* (Vol. 75, Issue 5). DOI: 10.4269/ajtmh.2006.75.973

Grishina, A., Chinnici, M., Kor, A.-L., Rondeau, E., Georges, J.-P., & De Chiara, D. (2021). *Data Center for Smart Cities: Energy and Sustainability Issue*. 1–36. DOI: 10.1007/978-3-030-38836-2_1

Grothmann, T., & Patt, A. (2005). Adaptive capacity and human cognition: The process of individual adaptation to climate change. *Global Environmental Change, 15*(3), 199–213. DOI: 10.1016/j.gloenvcha.2005.01.002

Gudoshava, M., Wanzala, M., Thompson, E., Mwesigwa, J., Endris, H. S., Segele, Z., Hirons, L., Kipkogei, O., Mumbua, C., Njoka, W., Baraibar, M., de Andrade, F., Woolnough, S., Atheru, Z., & Artan, G. (2022). Application of real time S2S forecasts over Eastern Africa in the co-production of climate services. *Climate Services, 27*, 100319. Advance online publication. DOI: 10.1016/j.cliser.2022.100319

Gunaseelan, R., Shanthi, S., Sowmya, R., & Datta, M. (2007). Areca nut use among rural residents of Sriperambudur Taluk: A qualitative study. *Indian Journal of Dental Research, 18*(1), 11–14. DOI: 10.4103/0970-9290.30915 PMID: 17347538

Guo, P., Li, Q., Guo, H., Li, H., & Yang, L. (2021). A bibliometric and visual analysis of global urban resilience research in 2011–2020: Development and hotspots. *Sustainability (Basel), 14*(1), 229. DOI: 10.3390/su14010229

Gupta, P. C., & Warnakulasuriya, S. (2002). Global epidemiology of areca nut usage. *Addiction Biology, 7*(1), 77–83. DOI: 10.1080/13556210020091437 PMID: 11900626

Gutierrez, K. S., & LePrevost, C. E. (2016). Climate justice in rural southeastern United States: A review of climate change impacts and effects on human health. *International Journal of Environmental Research and Public Health, 13*(2), 189. Advance online publication. DOI: 10.3390/ijerph13020189 PMID: 26848673

Hafiz Iqbal, M. (2020). Valuing ecosystem services of Sundarbans mangrove forest for improved conservation: Approach of randomized conjoint experiment. *Forestry Economics Review, 2*(1), 2631–3030. DOI: 10.1108/FER-04-2020-0008

Hajli, N. (2015). Social commerce constructs and consumer's intention to buy. *International Journal of Information Management, 35*(2), 183–191. DOI: 10.1016/j.ijinfomgt.2014.12.005

Halewood, M., López Noriega, I., & Louafi, S. (Eds.). (2013). *Crop Genetic Resources as a Global Commons: Challenges in International Law and Governance*. Routledge.

Hamman, P. (2017). *Definitions and redefinitions of urban sustainability: A bibliometric approach* (Vol. 11). Environnement Urbain/Urban Environment.

Hao, H., Wang, Y., & Chen, J. (2024). Empowering Scenario Planning with Artificial Intelligence: A Perspective on Building Smart and Resilient Cities. *Engineering (Beijing)*, *43*, 272–283. Advance online publication. DOI: 10.1016/j.eng.2024.06.012

Hashem, I. A. T., Usmani, R. S. A., Almutairi, M. S., Ibrahim, A. O., Zakari, A., Alotaibi, F., Alhashmi, S. M., & Chiroma, H. (2023). Urban Computing for Sustainable Smart Cities: Recent Advances, Taxonomy, and Open Research Challenges. *Sustainability 2023, Vol. 15, Page 3916, 15*(5), 3916. DOI: 10.3390/su15053916

He, K., Zhang, X., Ren, S., & Sun, J. (2016). Deep Residual Learning for Image Recognition.

Healy, L. M. (2008). *International Social Work: Professional Action in an Interdependent World* (2nd ed.). Oxford University Press., Available at https://global.oup.com/ushe/product/international-social-work-9780190922252?cc=in&lang=

Hiran, K. K., Khazanchi, D., Vyas, A. K., & Padmanaban, S. (Eds.). (2021). *Machine learning for sustainable development* (Vol. 9). Walter de Gruyter GmbH & Co KG., DOI: 10.1515/9783110702514

Hiwasaki, L., Luna, E., Syamsidik, , & Shaw, R. (2014). Process for integrating local and indigenous knowledge with science for hydro-meteorological disaster risk reduction and climate change adaptation in coastal and small island communities. *International Journal of Disaster Risk Reduction*, *10*, 15–27. DOI: 10.1016/j.ijdrr.2014.07.007

Hocevar, M., & Bartol, T. (2021). Mapping urban tourism issues: Analysis of research perspectives through the lens of network visualization. *International Journal of Tourism Cities*, *7*(3), 818–844. DOI: 10.1108/IJTC-05-2020-0110

Honadle, G. (1999). *How Context Matters: Linking Environmental Policy to People and Place*. Kumarian Press., https://api.semanticscholar.org/CorpusID:152578133

Hong, Y., Liu, Q., & Hu, X. (2021). AI technologies for disaster management. *International Journal of Disaster Risk Reduction*, *64*, 102495. DOI: 10.1016/j.ijdrr.2021.102495

Horton, D. (1998). Disciplinary roots and branches of evaluation: Some lessons from agricultural research. *Knowledge in Society*, *10*(4), 31–66. DOI: 10.1007/BF02912498

Hosseinian-Far, A., Ramachandran, M., & Slack, C. L. (2017). Emerging trends in cloud computing, big data, fog computing, IoT and smart living. In *Technology for Smart Futures* (pp. 29–40). Springer International Publishing., DOI: 10.1007/978-3-319-60137-3_2

Howard, P. H. (2009). Visualizing Consolidation in the Global Seed Industry: 1996–2008. *Sustainability (Basel)*, *1*(4), 1266–1287. DOI: 10.3390/su1041266

Huang, L., Bai, L., Zhang, X., & Gong, S. (2019). Re-understanding the antecedents of functional foods purchase: Mediating effect of purchase attitude and moderating effect of food neophobia. *Food Quality and Preference*, *73*, 266–275. DOI: 10.1016/j.foodqual.2018.11.001

Hurlimann, A. C., & March, A. P. (2012). The role of spatial planning in adapting to climate change. *Wiley Interdisciplinary Reviews: Climate Change*, *3*(5), 477–488. DOI: 10.1002/wcc.183

Hussain, I. (2024). Secure, Sustainable Smart Cities and the Internet of Things: Perspectives, Challenges, and Future Directions. *Sustainability 2024, Vol. 16, Page 1390*, *16*(4), 1390. DOI: 10.3390/su16041390

Ibrahim, B., & Mensah, H. (2022). Rethinking climate migration in sub-Saharan Africa from the perspective of tripartite drivers of climate change. *SN Social Sciences*, *2*(6), 87. Advance online publication. DOI: 10.1007/s43545-022-00383-y

İnce, E. C. (2025). Mapping the Path to Sustainable Urban Mobility: A Bibliometric Analysis of Global Trends and Innovations in Transportation Research. *Sustainability (2071-1050)*, *17*(4).

Intergovernmental Panel on Climate Change (IPCC). (2023). *Climate Change 2023: Synthesis Report*. Cambridge University Press., DOI: 10.1017/9781009325844

IPCC. (2022). Climate Change 2022: Impacts, Adaptation and Vulnerability. Contribution of Working Group II to the Sixth Assessment Report of the Intergovernmental Panel on Climate Change. Cambridge University Press.

IPES-Food. (2016). *From uniformity to diversity: A paradigm shifts from industrial agriculture to diversified agroecological systems*. International Panel of Experts on Sustainable Food Systems.

Islam, S. N., & Winkel, J. (2017). Climate change and social inequality. *DESA Working Paper No. 152*. https://doi.org/DOI: 10.18356/9dbcb5b4-en

Islam, S. T., Dey, A., Sanyal, N., & Bandyopadhyay, S. (2022). *Troubles in The Sundarbans: A study of social and ecological issues in Hingalganj Block, North 24 Parganas West Bengal*. ActionAid International.

Ismagilova, E., Slade, E. L., Rana, N. P., & Dwivedi, Y. K. (2020). The effect of electronic word of mouth communications on intention to buy: A meta-analysis. *Information Systems Frontiers*, *22*(5), 1203–1226. DOI: 10.1007/s10796-019-09924-y

Jabeen, H., Johnson, C., & Allen, A. (2010). Built-in resilience: Learning from grassroots coping strategies for climate variability. *Environment and Urbanization*, *22*(2), 415–431. DOI: 10.1177/0956247810379937

Jahan, R. (1995). *The elusive agenda: Mainstreaming women in development*. London: ZED Books. https://www.jstor.org/stable/41260001

Jalais, A. (2007). The Sundarbans: Whose World Heritage Site? *Conservation & Society*, *5*(3), 335–342. https://www.jstor.org/stable/26392892

Jamal, S., Ghosh, A., Hazarika, R., & Sen, A. (2022, July). Livelihood, conflict and tourism: An assessment of livelihood impact in Sundarbans, West Bengal. *International Journal of Geoheritage and Parks*, *10*(1), 383–399. Advance online publication. DOI: 10.1016/j.ijgeop.2022.07.004

Janik, A., Ryszko, A., & Szafraniec, M. (2020). Scientific landscape of smart and sustainable cities literature: A bibliometric analysis. *Sustainability (Basel)*, *12*(3), 779. DOI: 10.3390/su12030779

Jarvis, D. I., Hodgkin, T., Sthapit, B. R., Fadda, C., & Lopez-Noriega, I. (2016). A heuristic framework for identifying multiple ways of supporting the conservation and use of traditional crop varieties within the agricultural production system. *Critical Reviews in Plant Sciences*, *30*(1–2), 125–176.

Jasanoff, S., Hurlbut, J. B., & Saha, K. (2021). CRISPR democracy: Gene editing and the need for inclusive deliberation. *Issues in Science and Technology*, *37*(2), 25–32.

Javaid, S., Sufian, A., Pervaiz, S., & Tanveer, M. (2018). Smart traffic management system using Internet of Things. *International Conference on Advanced Communication Technology, ICACT, 2018-February*, 393–398. DOI: 10.23919/ICACT.2018.8323770

Jefferson, O. A., Köllhofer, D., Ehrich, T. H., & Jefferson, R. A. (2020). Transparency tools in gene patenting for informing policy and practice. *Nature Biotechnology*, *38*(4), 405–409. PMID: 24316644

Jewell, C. (2015). Who benefits from IP rights in agricultural innovation. *WIPO Article, August.*

Jia, B., Chen, Y., & Wu, J. (2021). Bibliometric analysis and research trend forecast of healthy urban planning for 40 years (1981–2020). *International Journal of Environmental Research and Public Health, 18*(18), 9444. DOI: 10.3390/ijerph18189444 PMID: 34574368

Jobin, A., Ienca, M., & Vayena, E. (2019). The global landscape of AI ethics guidelines. *Nature Machine Intelligence, 1*(9), 389–399. DOI: 10.1038/s42256-019-0088-2

Jongman, B., Ward, P. J., & Aerts, J. C. J. H. (2012). Global exposure to river and coastal flooding: Long term trends and changes. *Global Environmental Change, 22*(4), 823–835. Advance online publication. DOI: 10.1016/j.gloenvcha.2012.07.004

Joshua, M. D., Tompkins, E., Schreckenberg, K., Ngongondo, C., Gondwe, E., & Chiotha, S. (2022). Water policy and resilience of potable water infrastructure to climate risks in rural Malawi. *Physics and Chemistry of the Earth Parts A/B/C, 127*, 103155. Advance online publication. DOI: 10.1016/j.pce.2022.103155

Kamrowska-Załuska, D., & Bibri, S. E. (2021). Impact of AI-Based Tools and Urban Big Data Analytics on the Design and Planning of Cities. *Land 2021, Vol. 10, Page 1209, 10*(11), 1209. DOI: 10.3390/land10111209

Kanjilal, B., Bose, S., Patra, N., Barman, D., Ghosh, U., Mandal, A., Vardrevu, L., & Sengupta, P. (2013). *How Healthy are the Children of Indian Sundarbans?*http://r4d.dfid.gov.uk/pdf/outputs/FutureHealth_RPC/SHWforweb.pdf

Kanjilal, B., Mazumdar, P. G., Mukherjee, M., Mondal, S., Barman, D., & Singh, S. M. (2010, January). *Health care in the Sundarbans (India): Challenges. Future Health Systems Research Programme*, 21-30. Retrieved from https://assets.publishing.service.gov.uk/media/57a08b24ed915d3cfd000b68/sundarbans.pdf

Kanjilal, B., Mazumdar, P., Mukherjee, M., Mondal, S., Barman, D., Singh, S., & Mandal, A. (2010). Health care in the Sundarbans (India): Challenges and plan for a better future. *Future Health Systems Research Programme* (Issues 1–91). DOI: 10.6133/apjcn.012016.01

Keerthana, T., Kaviya, K., Deepthi Priya, S., & Kumar, A. S. (2021). AI enabled smart surveillance system. *Journal of Physics: Conference Series, 1916*(1), 12034. DOI: 10.1088/1742-6596/1916/1/012274

Kemp, S. P., Whittaker, J. K., & Tracy, E. M. (1997) *Person–Environment Practice: The Social Ecology of Interpersonal Helping*. New York: Aldine De Gruyter. https://api.semanticscholar.org/CorpusID:141687227

Kernaghan, S., & da Silva, J. (2013). Initiating and sustaining action: Experiences building resilience to climate change in Asian cities. *Urban Climate*, *7*, 47–63. DOI: 10.1016/j.uclim.2013.10.008

Khan, S. A. R., Umar, M., Yu, Z., & Nawaz, M. T. (2023). A Recent Digitalization in Recycling Industry Attaining Ecological Sustainability: A Comprehensive Outlook and Future Trend. *Environmental Science and Pollution Research International*, *30*(47), 103760–103775. DOI: 10.1007/s11356-023-29537-y PMID: 37695483

Khediri, A., Laouar, M. R., & Eom, S. B. (2021). Improving intelligent decision making in urban planning: Using machine learning algorithms. [IJBAN]. *International Journal of Business Analytics*, *8*(3), 40–58.

Kiewiet, B. (2005). Plant variety protection in the European Community. *World Patent Information*, *27*(4), 319–327. DOI: 10.1016/j.wpi.2005.07.006

Kingsbury, N. (2009). *Hybrid: The History and Science of Plant Breeding*. University of Chicago Press. DOI: 10.7208/chicago/9780226437057.001.0001

Kloppenburg, J. (2004). *First the Seed: The Political Economy of Plant Biotechnology, 1492–2000* (2nd ed.). University of Wisconsin Press.

Kloppenburg, J. (2010). Impeding Dispossession, Enabling Repossession: Biological Open Source and the Recovery of Seed Sovereignty. *Journal of Agrarian Change*, *10*(3), 367–388. DOI: 10.1111/j.1471-0366.2010.00275.x

Kobayashi, T., Tanaka, K., & Kobayashi, K. (2019). Wildfire detection from satellite images using deep convolutional neural networks. *Remote Sensing*, *11*(23), 2874. DOI: 10.3390/rs11232874

Korada, L. (2021). Unlocking Urban Futures: The Role Of Big Data Analytics And AI In Urban Planning – A Systematic Literature Review And Bibliometric Insight. *Migration Letters : An International Journal of Migration Studies*, *18*(6), 775–795. https://migrationletters.com/index.php/ml/article/view/11260

Köster, F., Ulmer, M. W., Mattfeld, D. C., & Hasle, G. (2018). Anticipating emission-sensitive traffic management strategies for dynamic delivery routing. *Transportation Research Part D, Transport and Environment*, *62*, 345–361. DOI: 10.1016/j.trd.2018.03.002

Koumetio Tekouabou, S. C., Diop, E. B., Azmi, R., & Chenal, J. (2022). Artificial Intelligence Based Methods for Smart and Sustainable Urban Planning: A Systematic Survey. *Archives of Computational Methods in Engineering*, *30*(2), 1421–1438. DOI: 10.1007/s11831-022-09844-2

Kourtit, K. (2021). City intelligence for enhancing urban performance value: A conceptual study on data decomposition in smart cities. *Asia-Pacific Journal of Regional Science*, *5*(1), 191–222. DOI: 10.1007/s41685-021-00193-9

Krabokoukis, T., & Polyzos, S. (2023). A bibliometric analysis of integrating tourism development into urban planning. *Sustainability (Basel)*, *15*(20), 14886. DOI: 10.3390/su152014886

Kreibich, H., Bubeck, P., Van Vliet, M., & De Moel, H. (2017). A review of damage-reducing measures to manage fluvial flood risks in a changing climate. *Mitigation and Adaptation Strategies for Global Change*, *22*(5), 963–989. DOI: 10.1007/s11027-016-9727-5

La Barbera, F., & Ajzen, I. (2020). Control interactions in the theory of planned behavior: Rethinking the role of subjective norm. *Europe's Journal of Psychology*, *16*(3), 401–417. DOI: 10.5964/ejop.v16i3.2056 PMID: 33680190

La Vía Campesina. (2015). *Seeds: Commons or corporate property?*https://viacampesina.org

Laird, S. A., & Wynberg, R. P. (2018). *A fact-finding and scoping study on digital sequence information on genetic resources in the context of the Convention on Biological Diversity and the Nagoya Protocol*. Secretariat of the Convention on Biological Diversity.

Land Transport Authority (LTA). (2025). Intelligent Transport Systems. https://www.lta.gov.sg/content/ltagov/en/getting_around/driving_in_singapore/intelligent_transport_systems.html

Lazar, N., & Chithra, K. (2021). Comprehensive bibliometric mapping of publication trends in the development of Building Sustainability Assessment Systems. *Environment, Development and Sustainability*, *23*(4), 4899–4923. DOI: 10.1007/s10668-020-00796-w

Leauthaud, C., Duvail, S., Hamerlynck, O., Paul, J. L., Cochet, H., Nyunja, J., Albergel, J., & Grünberger, O. (2013). Floods and livelihoods: The impact of changing water resources on wetland agro-ecological production systems in the Tana River Delta, Kenya. *Global Environmental Change*, *23*(1). Advance online publication. DOI: 10.1016/j.gloenvcha.2012.09.003

Leck, H., & Simon, D. (2012). Fostering multiscalar collaboration and co-operation for effective governance of climate change adaptation. *Urban Studies (Edinburgh, Scotland)*, *50*(6), 1221–1238. Advance online publication. DOI: 10.1177/0042098012461675

Leichenko, R. (2011). Climate change and urban resilience. *Current Opinion in Environmental Sustainability*, *3*(3), 164–168. DOI: 10.1016/j.cosust.2010.12.014

Lemaire, X. (2021). Climate change and strategic low-carbon planning in African cities after COVID-19: Inclusiveness or chaos? *Journal of the British Academy*, *9*, 39–79. Advance online publication. DOI: 10.5871/jba/009s9.039

Leskien, D., & Flitner, M. (1997). *Intellectual Property Rights and Plant Genetic Resources: Options for a Sui Generis System*. IPGRI.

Liang, D., De Jong, M., Schraven, D., & Wang, L. (2022). Mapping key features and dimensions of the inclusive city: A systematic bibliometric analysis and literature study. *International Journal of Sustainable Development and World Ecology*, *29*(1), 60–79. DOI: 10.1080/13504509.2021.1911873

Liao, P., Wan, Y., Tang, P., Wu, C., Hu, Y., & Zhang, S. (2019). Applying crowdsourcing techniques in urban planning: A bibliometric analysis of research and practice prospects. *Cities (London, England)*, *94*, 33–43. DOI: 10.1016/j.cities.2019.05.024

Li, C., Yu, W., Dzodzomenyo, M., Asamoah, M., Kerapetse, C. T., Kandel, M., & Wright, J. (2021). Growing Spatial Overlap Between Dam-Related Flooding, Cropland and Domestic Water Points: A Water–Energy–Food Nexus Management Challenge in Malawi and Ghana. *Frontiers in Water*, *3*, 730370. Advance online publication. DOI: 10.3389/frwa.2021.730370

Li, H., Xiang, Y., Yang, W., Lin, T., Xiao, Q., & Zhang, G. (2024). Green roof development knowledge map: A review of visual analysis using CiteSpace and VOSviewer. *Heliyon*, *10*(3), e24958. DOI: 10.1016/j.heliyon.2024.e24958 PMID: 38356566

Li, J., Meng, Y., Ma, L., Du, S., Zhu, H., Pei, Q., & Shen, X. (2022). A Federated Learning Based Privacy-Preserving Smart Healthcare System. *IEEE Transactions on Industrial Informatics*, *18*(3), 2021–2031. DOI: 10.1109/TII.2021.3098010

Li, S., Yuan, F., & Liu, J. (2024). Smart city VR landscape planning and user virtual entertainment experience based on artificial intelligence. *Entertainment Computing*, *51*, 100743. DOI: 10.1016/j.entcom.2024.100743

Liu, Y., Liu, X., & Li, Y. (2022). Advances in AI-powered disaster risk reduction: A review. *International Journal of Disaster Risk Science*, *13*(1), 1–18. DOI: 10.1007/s13753-022-00379-x

Liu, Y., & Wang, Z. (2020). Machine Learning for Predictive Modeling of Geospatial Data: A Review. *Computers, Environment and Urban Systems*, *82*, 101489.

Liu, Y., & Xie, L. (2021). Machine Learning in Geospatial Data Processing: A Review of Algorithms and Applications. *Journal of Geographic Information Science, 35*(2), 145–162.

Li, Z., Al Hassan, R., Shahidehpour, M., Bahramirad, S., & Khodaei, A. (2019). A Hierarchical Framework for Intelligent Traffic Management in Smart Cities. *IEEE Transactions on Smart Grid, 10*(1), 691–701. DOI: 10.1109/TSG.2017.2750542

Llewelyn, M., & Adcock, M. (2006). *European plant intellectual property*. Bloomsbury Publishing.

Long, J., Shelhamer, E., & Darrell, T. (2015). Fully Convolutional Networks for Semantic Segmentation. *Proceedings of the IEEE Conference on Computer Vision and Pattern Recognition (CVPR),* 3431-3440.

Louafi, S. (2020). Governing digital sequence information to ensure fairness in the use of genetic resources. *Frontiers in Plant Science, 11*, 628512. DOI: 10.3389/fpls.2020.628512

Louafi, S. (2022). Digital sequence information and the conservation and sustainable use of biodiversity. *Frontiers in Genetics, 13*, 808459.

Louwaars, N. P. (2007). *Seeds of Confusion: The Impact of Policies on Seed Systems*. Wageningen University.

Louwaars, N. P., Dons, H., Overwalle, G. V., Raven, H., Arundel, A., Eaton, D., & Nelis, A. (2005). *Breeding Business: The Future of Plant Breeding in the Light of Developments in Patent Rights and Plant Breeders' Rights*. Centre for Genetic Resources.

Louwaars, N., & De Boef, W. (2012). Integrated Seed Sector Development in Africa: A Conceptual Framework for Creating Coherence between Practices, Programs, and Policies. *Journal of Crop Improvement, 26*(1), 39–59. DOI: 10.1080/15427528.2011.611277

Lv, T., Wang, L., Xie, H., Zhang, X., & Zhang, Y. (2021). Exploring the global research trends of land use planning based on a bibliometric analysis: Current status and future prospects. *Land (Basel), 10*(3), 304. DOI: 10.3390/land10030304

Maas, P., Iyer, S., Gros, A., Park, W., & McGorman, L. (2020). Facebook Disaster Maps: Aggregate insights for crisis response & recovery. *Proceedings of the 27th ACM SIGSPATIAL International Conference on Advances in Geographic Information Systems*, 579–582. https://doi.org/DOI: 10.1145/3347146.3359332

Mahadevia, K., & Vikas, M. (2012). Climate change—Impact on the Sunderbans: A case study. *International Scientific Journal-Environmental Science*, *2*(1), 7–15.

Mahmoud, M., Perez-Ortiz, M., Asad Bokhari, S. A., & Myeong, S. (2022). Use of Artificial Intelligence in Smart Cities for Smart Decision-Making: A Social Innovation Perspective. *Sustainability 2022, Vol. 14, Page 620, 14*(2), 620. DOI: 10.3390/su14020620

Majeed, K., Jing, Y., Novakovic, D., & Ouazzane, P. K. (2014). Behaviour Based Anomaly Detection for Smartphones Using Machine Learning Algorithm. *International Conference on Computer Science and Information Systems.*

Mangani, R., Tesfamariam, E. H., Bellocchi, G., & Hassen, A. (2018). Growth, development, leaf gaseous exchange, and grain yield response of maize cultivars to drought and flooding stress. *Sustainability (Basel)*, *10*(10), 3492. Advance online publication. DOI: 10.3390/su10103492

Marvuglia, A., Havinga, L., Heidrich, O., Fonseca, J., Gaitani, N., & Reckien, D. (2020). Advances and challenges in assessing urban sustainability: An advanced bibliometric review. *Renewable & Sustainable Energy Reviews*, *124*, 109788. DOI: 10.1016/j.rser.2020.109788

Matin, N., Forrester, J., & Ensor, J. (2018). What is equitable resilience? *World Development*, *109*, 197–205. DOI: 10.1016/j.worlddev.2018.04.020 PMID: 30177865

Mehmood, I., & Ning, H. (2024). Role of computer vision in smart cities: applications and research challenges. *Multimedia Tools and Applications 2024 83:5, 83*(5), 14883–14883. DOI: 10.1007/s11042-024-18276-y

Mehrabi, N., Morstatter, F., Saxena, N., Lerman, K., & Galstyan, A. (2021). A survey on bias and fairness in machine learning. *ACM Computing Surveys*, *54*(6), 1–35. DOI: 10.1145/3457607

Mehrtash, H., Duncan, K., Parascandola, M., David, A., Gritz, E. R., Gupta, P. C., Mehrotra, R., Amer Nordin, A. S., Pearlman, P. C., Warnakulasuriya, S., Wen, C.-P., Zain, R. B., & Trimble, E. L. (2017). Defining a global research and policy agenda for betel quid and areca nut. *The Lancet. Oncology*, *18*(12), e767–e775. DOI: 10.1016/S1470-2045(17)30460-6 PMID: 29208442

Mertz, O., Mbow, C., Reenberg, A., & Diouf, A. (2009). Farmers' perceptions of climate change and agricultural adaptation strategies in rural Sahel. *Environmental Management*, *43*(5), 804–816. DOI: 10.1007/s00267-008-9197-0 PMID: 18810526

Midgley, J. (1994). Defining Social Development: Historical Trends and Conceptual Formulations. *Social Development Issues*, *16*(3), 3–19. DOI: 10.4135/9781446221839.n2

Migliori, I. K. (2021). The current status of plant-related IP Rights in Brazil: How to reap the fruits of the protection culture in the agribusiness field. *Stockholm Intellectual Property Law Review*, (1), 48–55.

Miller, S. E., Hayward, R. A., & Shaw, T. V. (2012). Environmental Shifts for Social Work: A Principles Approach. *International Journal of Social Welfare*, *21*(3), 270–277. DOI: 10.1111/j.1468-2397.2011.00848.x

Mimi, R., & Haynes, L. (2018). *The role of livelihood activities and income portfolios in shaping women's perceptions of wellbeing and adaptation choices, in the Indian Bengal Delta Statement of Originality.*

Mishra, M., Acharyya, T., Santos, C. A. G., da Silva, R. M., Kar, D., Mustafa Kamal, A. H., & Raulo, S. (2021). Geo-ecological impact assessment of severe cyclonic storm Amphan on Sundarban mangrove forest using geospatial technology. *Estuarine, Coastal and Shelf Science*, *260*, 107486. DOI: 10.1016/j.ecss.2021.107486

Mistri, A. (2019). Is the migration from Indian Sundarban an environmental migration? Investigating through Sustainable Livelihood Approach (SLA). *Asian Profile*, *47*(3), 195–218.

Mitlin, D., & Satterthwaite, D. (2012). Addressing poverty and inequality new forms of urban governance in Asia. *Environment and Urbanization*, *24*(2), 395–401. DOI: 10.1177/0956247812458062

Mitra, R. (2024). *Nature-based Solutions can Help Sundarbans Survive. DownToEarth.* Retrieved from https://www.downtoearth.org.in/news/climate-change/nature-based-solutions-can-help-sundarbans -survive-95230

Mizanur Rahman, M., Rakib Hossain, M., & Nazrul Islam, M. (2021). Climate Change Impact on Sundarbans: Challenges for Mitigation Strategies. *Climate Change in Bangladesh: A Cross-Disciplinary Framework*, 47-64.

Mnih, V., Kavukcuoglu, K., Silver, D., Rusu, A. A., Veness, J., Bellemare, M. G., Graves, A., Riedmiller, M., Fidjeland, A. K., Ostrovski, G., Petersen, S., Beattie, C., Sadik, A., Antonoglou, I., King, H., Kumaran, D., Wierstra, D., Legg, S., & Hassabis, D. (2015). Human-level control through deep reinforcement learning. *Nature*, *518*(7540), 529–533. DOI: 10.1038/nature14236 PMID: 25719670

Moeller, N. I., & Stannard, C. (2013). Identifying benefit flows. Studies on the Potential Monetary and Non-Monetary Benefits Arising from the International Treaty on Plant Genetic Resources for Food and Agriculture, FAO-Treaty Secretariat.

Molaei Qelichi, M., Ghasemi, K., Tepe, E., & Murgante, B. (2025). Mapping Trends in Urban Livability Research: A Comprehensive Bibliometric Analysis. *Journal of Planning Literature*, •••, 08854122241312181. DOI: 10.1177/08854122241312181

Molinari, N. (2017). Intensifying Insecurities: The impact of climate change on vulnerability to human trafficking in the Indian Sundarbans. *Anti Trafficking Review*, *363*(8), 1–11. DOI: 10.14197/atr.20121784

Møller-Jensen, L., Agergaard, J., Andreasen, M. H., Kofie, R. Y., Yiran, G. A. B., & Oteng-Ababio, M. (2023). Probing political paradox: Urban expansion, floods risk vulnerability and social justice in urban Africa. *Journal of Urban Affairs*, *45*(3), 505–521. Advance online publication. DOI: 10.1080/07352166.2022.2108436

Mondal, M. K., Bhuiyan, S. I., & Franco, D. T. (2001). Soil salinity reduction and prediction of salt dynamics in the coastal rice lands of Bangladesh. *Agricultural Water Management*, *47*(1), 9–23. DOI: 10.1016/S0378-3774(00)00098-6

Mondal, T., Bowers, W. W., & Hossen, M. S. (2021, March 25). Sustainable Management of Sundarbans: Stakeholder Attitudes Towards Participatory Management and Conservation of Mangrove Forests. *Journal of Sustainable Development*, *14*(3), 23–40. DOI: 10.5539/jsd.v14n3p23

Montoliu, L. (2019). The CRISPR revolution: Impact on genomics, biotechnology and society. *EMBO Reports*, *20*(9), e48595. DOI: 10.15252/embr.201948595

Mooney, P. R. (2011). The law of the seed: Another development and plant genetic resources. *Development Dialogue*, *1*(1), 1–75.

Moon, J., Chadee, D., & Tikoo, S. (2008). Culture, product type, and price influences on consumer purchase intention to buy personalized products online. *Journal of Business Research*, *61*(1), 31–39. DOI: 10.1016/j.jbusres.2006.05.012

Moore, J. (2023, March 6). *PayBitoPro saves Sundarbans' child education*. Retrieved from PayBitoPro: https://www.paybito.com/lost-in -the -mangroves -paybito -steps -up -to -save -child -education -in -sundarbans/ 1

Mori, J., Ohta, Y., & Saito, T. (2019). Tsunami early warning and real-time monitoring using big data and AI technologies. *Pure and Applied Geophysics*, *176*(5), 2021–2036. DOI: 10.1007/s00024-018-2010-1

Morley, J., Floridi, L., Kinsey, L., & Elhalal, A. (2021). From what to how: An initial review of publicly available AI ethics tools, methods and research to translate principles into practices. *Science and Engineering Ethics*, *27*(1), 1–19. DOI: 10.1007/s11948-021-00282-w PMID: 33492516

Mortaheb, R., & Jankowski, P. (2023). Smart city re-imagined: City planning and GeoAI in the age of big data. *Journal of Urban Management*, *12*(1), 4–15. DOI: 10.1016/j.jum.2022.08.001

Mosavi, A., Ozturk, P., & Chau, K. W. (2018). Flood prediction using machine learning models: Literature review. *Water (Basel)*, *10*(11), 1536. DOI: 10.3390/w10111536

Moser, S. C., & Ekstrom, J. A. (2010). A framework to diagnose barriers to climate change adaptation. *Proceedings of the National Academy of Sciences of the United States of America*, *107*(51), 22026–22031. DOI: 10.1073/pnas.1007887107 PMID: 21135232

Mozumdar, M., Shamsuzzaman, M., Un-Nabi, R., & Harun-Al-Rashid, A. (2018). Socio-economic characteristics and fishing operation activities of the artisanal fishers in the Sundarbans Mangrove Forest, Bangladesh. *Turkish Journal of Fisheries and Aquatic Sciences*, *18*(6), 788–799. DOI: 10.4194/1303-2712-v18_6_05

Mubangizi, A., Wanyama, J., Kiggundu, N., & Nakawuka, P. (2023). Assessing Suitability of Irrigation Scheduling Decision Support Systems for Lowland Rice Farmers in Sub-Saharan Africa—A Review. *Agricultural Sciences*, *14*(02), 219–239. Advance online publication. DOI: 10.4236/as.2023.142015

Mukhopadhyay, A. (2009). *Cyclone Aila and the Sundarbans: An Enquiry into the Disaster and Politics of Aid and Relief.* In Research Report on CRG Annual Winter Course on Forced Migration.

Müller, A., & Jain, R. (2020). Enhancing Geospatial Data Integration with AI/ML Techniques. *International Journal of AI and Data Mining*, *9*(1), 22–39.

Müller, C., Ouédraogo, W. A., Schwarz, M., Barteit, S., & Sauerborn, R. (2023). The effects of climate change-induced flooding on harvest failure in Burkina Faso: Case study. *Frontiers in Public Health*, *11*, 1166913. Advance online publication. DOI: 10.3389/fpubh.2023.1166913 PMID: 37614457

Müller, C., Waha, K., Bondeau, A., & Heinke, J. (2014). Hotspots of climate change impacts in sub-Saharan Africa and implications for adaptation and development. *Global Change Biology*, *20*(8), 2505–2517. Advance online publication. DOI: 10.1111/gcb.12586 PMID: 24796720

Mulligan, C., Olivadoti, L., & Buck, C. (2020). Edge AI: The convergence of AI and edge computing. *Journal of Information Technology Teaching Cases*, *10*(1), 31–36. DOI: 10.1177/2043886920902736

Nagata, J. M., Hampshire, K., Epstein, A., Lin, F., Zakaras, J., Murnane, P., Charlebois, E. D., Tsai, A. C., Nash, D., & Weiser, S. D. (2022). Analysis of Heavy Rainfall in Sub-Saharan Africa and HIV Transmission Risk, HIV Prevalence, and Sexually Transmitted Infections, 2005-2017. *JAMA Network Open*, *5*(9), e2230282. Advance online publication. DOI: 10.1001/jamanetworkopen.2022.30282 PMID: 36074468

Narayanan, A., & Shmatikov, V. (2008). Robust de-anonymization of large sparse datasets. *IEEE Symposium on Security and Privacy*, 111–125. DOI: 10.1109/SP.2008.33

Neelakandan, S., Berlin, M. A., Tripathi, S., Devi, V. B., Bhardwaj, I., & Arulkumar, N. (2021). IoT-based traffic prediction and traffic signal control system for smart city. *Soft Computing*, *25*(18), 12241–12248. DOI: 10.1007/s00500-021-05896-x

Negev, M., Teschner, N., Rosenthal, A., Levine, H., Lew-Levy, C., & Davidovitch, N. (2019). Adaptation of health systems to climate-related migration in Sub-Saharan Africa: Closing the gap. *International Journal of Hygiene and Environmental Health*, *222*(2), 311–314. Advance online publication. DOI: 10.1016/j.ijheh.2018.10.004 PMID: 30503929

Neumann, B., Vafeidis, A. T., Zimmermann, J., & Nicholls, R. J. (2015). Future coastal population growth and exposure to sea-level rise and coastal flooding - A global assessment. *PLoS One*, *10*(3), e0118571. Advance online publication. DOI: 10.1371/journal.pone.0118571 PMID: 25760037

New York City, D. O. T. (2025). Infrastructure: Traffic Signals. https://www.nyc.gov/html/dot/html/infrastructure/signals.shtml

Nian, N., Suk, H. I., & Kim, H. J. (2013). HOG feature enhancement for segmented handwritten character recognition.

Nicholls, R. J., Adger, W. N., Hutton, C. W., & Hanson, S. E. (2020). *Deltas in the Anthropocene*. Palgrave Macmillan Springer Nature., DOI: 10.1007/978-3-030-23517-8

Nijar, G. S. (2011). *The Nagoya Protocol on Access and Benefit Sharing: An Analysis*. Third World Network.

Njoku, N., Lamond, J., Everett, G., & Manu, P. (2015). An overview of municipal solid waste management in developing and developed economies: Analysis of practices and contributions to urban flooding in Sub-Saharan Africa. *12th International Postgraduate Research Conference Proceedings*.

Norgaard, K. M. (2017, September). The Sociological Imagination in a Time of Climate Change. *Global and Planetary Change, Elsevier*. DOI: DOI: 10.1016/j.gloplacha.2017.09.018

Norris, F., Stevens, S., Pfefferbaum, B., Wyche, K., & Pfefferbaum, R. (2008). Community resilience as metaphor, theory, set of capacities, and strategy for disaster readiness. *American Journal of Community Psychology*, *41*(1–2), 127–150. DOI: 10.1007/s10464-007-9156-6 PMID: 18157631

Nyong, A., Adesina, F., & Elasha, B. O. (2007). The value of indigenous knowledge in climate change mitigation and adaptation strategies in the African Sahel. *Mitigation and Adaptation Strategies for Global Change*, *12*(5), 787–797. DOI: 10.1007/s11027-007-9099-0

O'Brien, K. (2012). Global environmental change II: From adaptation to deliberate transformation. *Progress in Human Geography*, *36*(5), 667–676. DOI: 10.1177/0309132511425767

O'Brien, K., Eriksen, S., Nygaard, L. P., & Schjolden, A. (2007). Why different interpretations of vulnerability matter in climate change discourses. *Climate Policy*, *7*(1), 73–88. DOI: 10.1080/14693062.2007.9685639

O'Neil, C. (2016). *Weapons of math destruction: How big data increases inequality and threatens democracy*. Crown Publishing Group.

OECD. (2019). *OECD Principles on Artificial Intelligence*. Retrieved from https://www.oecd.org/going-digital/ai/principles/

Oguamanam, C. (2014). Farmers' rights and the intellectual property dynamic in agriculture. *Sage Handbook of Intellectual Property*, 238-257.

Okoye, K., Nganji, J. T., Hiran, K. K., & Hosseini, S. (2024, May). Impact and implications of AI methods and tools for the future of education. [). Frontiers Media SA.]. *Frontiers in Education*, *9*, 1434052. DOI: 10.3389/feduc.2024.1434052

Olhoff, A., & Schaer, C. (2010). *Screening tools and guidelines to support the mainstreaming of climate change adaptation into development assistance a stocktaking report*. New York: UNDP. http://www.undp.org/climatechange/library.shtml

Oliva, E. J. D., Sánchez-Torres, J. A., & Nino, J. S. S. (2022). City brand evolution research and future agenda: Bibliometric and science mapping analysis. *Heliyon*, *8*(8). PMID: 35991998

Omer, S. (1979). *'Social Development'. International Social Work*, *22*(3), 11–26. DOI: 10.1177/002087287902200303

Pal, P., & Poddar, S. (2024). Applications of Smart Farming Sensors: A Way Forward. *Data Science for Agricultural Innovation and Productivity*, 175.

Panda, S., Sadhu, C., Pramanik, G., Pahari, S., & Hossain, J. (2016). Concerning public health situation of under-nutrition in children and anemia in women in Indian Sundarbans delta: A community based cross sectional investigation. *BMC Nutrition*, *2*(1), 1–10. DOI: 10.1186/s40795-016-0105-3

Pascariu, G. C., Banica, A., & Nijkamp, P. (2023). A meta-overview and bibliometric analysis of resilience in spatial planning–the relevance of place-based approaches. *Applied Spatial Analysis and Policy*, *16*(3), 1097–1127. DOI: 10.1007/s12061-022-09449-z PMID: 35495415

Pasquier, T., Lau, M. K., Trippas, J. R., & Jay, C. (2020). Artificial intelligence for disaster response and resilience: Recent advances and future directions. *IEEE Security and Privacy*, *18*(4), 29–36. DOI: 10.1109/MSEC.2020.2998726

Patel, H., Jain, V., & Sharma, A. (2019). Operation Blue Sky: An AI-powered flood forecasting system for India. *IBM Journal of Research and Development*, *63*(6), 1–12. DOI: 10.1147/JRD.2019.2952187

Pato, M. L., & Duque, A. S. (2025). Mapping Innovation and Sustainability in Rural Tourism: A Bibliometric Approach. *Sustainability (Basel)*, *17*(6), 2574. DOI: 10.3390/su17062574

Pauchard, N., & Godt, C. (2021). Legal challenges of digital sequence information: Implications for access and benefit-sharing. *Journal of Environmental Law*, *33*(1), 1–25. DOI: 10.1093/jel/eqa

Peeters, J. (2012). The Place of Social Work in Sustainable Development: Towards Ecosocial Practice. *International Journal of Social Welfare*, *21*(3), 287–298. DOI: 10.1111/j.1468-2397.2011.00856.x

Pelling, M. (2003). *The vulnerability of cities: Natural disasters and social resilience*. Earthscan., DOI: 10.1080/15693430500111884

Pelling, M. (2011). *Adaptation to climate change: From resilience to transformation*. Routledge., DOI: 10.4324/9780203889046

Pelling, M., & Manuel-Navarrete, D. (2011). From resilience to transformation: The adaptive cycle in two Mexican urban centers. *Ecology and Society*, *16*(2), 11. DOI: 10.5751/ES-04038-160211

Perez-Saez, J., Lessler, J., Lee, E. C., Luquero, F. J., Malembaka, E. B., Finger, F., Langa, J. P., Yennan, S., Zaitchik, B., & Azman, A. S. (2022). The seasonality of cholera in sub-Saharan Africa: A statistical modelling study. *The Lancet. Global Health*, *10*(6), e831–e839. Advance online publication. DOI: 10.1016/S2214-109X(22)00007-9 PMID: 35461521

Perrow, C. (2011). *Normal accidents: Living with high-risk technologies* (2nd ed.). Princeton University Press. DOI: 10.2307/j.ctt7srgf

Pervin, M., Sultana, S., Phirum, A., Camara, I., Nzau, M., Phonnasane, V., & Anderson, S. (2013). *A framework for mainstreaming climate resilience into development planning*. IIED Working Paper, IIED, London https://www.iied.org/10050iied

Peschard, K. (2017). Seed wars and farmers' rights: Comparative perspectives from Brazil and India. *The Journal of Peasant Studies*, *44*(1), 144–168. DOI: 10.1080/03066150.2016.1191471

Peschard, K., & Randeria, S. (2020). 'Keeping seeds in our hands': Legal pluralism and seed governance in the Global South. *Third World Quarterly*, *41*(6), 1012–1030.

Peters, L., Thaler, T., Hanger-Kopp, S., & Seebauer, S. (2017). Understanding the dynamics of risk perception and community resilience. *Natural Hazards*, *86*(1), 353–376. DOI: 10.1007/s11069-016-2705-3

Pharmacogn, J., Srilakshmi, A., Rakkini, J., Sekar, K. R., & Manikandan, R. (2018). A Comparative study on Internet Of Things (IoT) and its applications in Smart Agriculture. *Pharmacognosy Journal*, *10*(2), 260–264. DOI: 10.5530/pj.2018.2.46

Pimbert, M. (2009). *Towards food sovereignty: Reclaiming autonomous food systems*. IIED.

Pitchaikani, J. S. (2020, March). Integrated coastal zone management practices for Sundarbans, India. *Indian Journal of Geo-Marine Sciences*, *49*(03), 352–356. https://core.ac.uk/download/pdf/326041666.pdf

Polin, F., & Alam, D. (2020, March 13). *Enhancing Community Participation in Forest Management: Case Study of the Sundarbans, Bangladesh*. South-West Region (Khulna Division). Retrieved from https://www.researchgate.net/publication/339916369_Enhancing_Community_Participation_in_Forest_Management_Case_Study_of_the_Sundarbans_Bangladesh

Posey, D. A., & Dutfield, G. (1996). *Beyond Intellectual Property: Toward Traditional Resource Rights for Indigenous Peoples and Local Communities*. IDRC.

Prakash, A. (2017). The Politics of Seed Laws in India: A Historical Institutional Analysis. *The Journal of Peasant Studies*, *44*(1), 1–22.

Pramanik, M., Szabo, S., Pal, I., Udmale, P., Pongsiri, M., & Chilton, S. (2021). Population health risks in multi-hazard environments: Action needed in the Cyclone Amphan and COVID-19 - hit Sundarbans region, India. *Climate and Development*, *14*(2), 99–104. DOI: 10.1080/17565529.2021.1889948

Prasad, P. V., Bhatnagar, N., Bhandari, V., Jacob, G., Narayan, K., Echeverría, R., Beintema, N., Farah Cox, P., & Compton, J. (2023). Patterns of investment in agricultural research and innovation for the Global South, with a focus on sustainable agricultural intensification. *Frontiers in Sustainable Food Systems*, *7*, 1108949. DOI: 10.3389/fsufs.2023.1108949

Prip, C. (2024). Farmers' Rights in the Plant Treaty: Interrelations and recent interactions with other international regimes and processes. *Development Studies Research*, *11*(1), 2357095. Advance online publication. DOI: 10.1080/21665095.2024.2357095

Qian, H., Zhang, F., & Qiu, B. (2023). Deciphering the evolution, frontier, and knowledge clustering in sustainable city planning: A 60-year interdisciplinary review. *Sustainability (Basel)*, *15*(24), 16854. DOI: 10.3390/su152416854

Qian, K., Mao, L., Liang, X., Ding, Y., Gao, J., Wei, X., Guo, Z., & Li, J. (2023). *AI Agent as Urban Planner: Steering Stakeholder Dynamics in Urban Planning via Consensus-based Multi-Agent Reinforcement Learning*. https://arxiv.org/abs/2310.16772v2

Qu, S., Li, H., Wu, J., & Zhao, B. (2025). Environmental and Health Benefits: A Bibliometric and Knowledge Mapping Analysis of Research Progress. *Sustainability (Basel)*, *17*(5), 2269. DOI: 10.3390/su17052269

Rabby, M. K. M., Islam, M. M., & Imon, S. M. (2019). A review of IoT application in a smart traffic management system. *2019 5th International Conference on Advances in Electrical Engineering, ICAEE 2019*, 280–285. DOI: 10.1109/ICAEE48663.2019.8975582

RAFI. (1998). *Patents, plants and plunder*. Rural Advancement Foundation International.

Rahman, M., Giedraitis, V. R., Lieberman, L. S., Akhtar, T., & Taminskienė, V. (2013). Shrimp cultivation with water salinity in Bangladesh: The implications of an ecological model. *Universal Journal of Public Health*, *1*(3), 131–142. DOI: 10.13189/ujph.2013.010313

Raj, K. (2007). *Relocating Modern Science: Circulation and the Construction of Knowledge in South Asia and Europe, 1650–1900*. Palgrave Macmillan. DOI: 10.1057/9780230625310

Ramanna, A. (2006). India's Plant Variety and Farmers' Rights Legislation: Potential Benefits and Harms. *The Journal of World Intellectual Property*, *9*(2), 189–210.

Ramiaramanana, F. N., & Teller, J. (2021). Urbanization and floods in sub-saharan africa: Spatiotemporal study and analysis of vulnerability factors—case of antananarivo agglomeration (Madagascar). *Water (Basel)*, *13*(2), 149. Advance online publication. DOI: 10.3390/w13020149

Ranyal, E., Sadhu, A., & Jain, K. (2022). Road Condition Monitoring Using Smart Sensing and Artificial Intelligence: A Review. *Sensors 2022, Vol. 22, Page 3044*, *22*(8), 3044. DOI: 10.3390/s22083044

Rathore, M. M., Paul, A., Ahmad, A., & Jeon, G. (2020). IoT-Based Big Data: From Smart City towards Next Generation Super City Planning. Https://Services.Igi-Global.Com/Resolvedoi/Resolve.Aspx?Doi=10.4018/978-1-5225-9866-4.Ch064, 1409–1428. DOI: 10.4018/978-1-5225-9866-4.ch064

Ray, P. S. (2020). Socio-Economic History in Sundarbans. *Journal of Critical Reviews*, *7*(2), 1030–1034.

Razavi, M., Hamidkhani, M., & Sadeghi, R. (2019). Smart Traffic Light Scheduling in Smart City Using Image and Video Processing. *Proceedings of 3rd International Conference on Internet of Things and Applications, IoT 2019*. DOI: 10.1109/IICITA.2019.8808836

Razia, S., & Ah, S. H. B. A. B. (2022). Panoramic Mapping of Urban Social Sustainability: A 35-Year Bibliometric and Visualization Analysis. *Journal of Regional and City Planning*, *33*(2), 49–78. DOI: 10.5614/jpwk.2022.33.2.4

Reid, H., Alam, M., Berger, R., Cannon, T., Huq, S., & Milligan, A. (2009). *Community-based adaptation to climate change: An overview*. Participatory Learning and Action: Community Based Adaptation to Climate Change, 60, 11–33. Available at https://www.researchgate.net/publication/272791313_Community-based_adaptation_to_climate_change_An_overview

Rentschler, J., Salhab, M., & Jafino, B. A. (2022). Flood exposure and poverty in 188 countries. *Nature Communications*, *13*(1), 3527. Advance online publication. DOI: 10.1038/s41467-022-30727-4 PMID: 35764617

Ren, Y., Leng, Y., Qi, J., Sharma, P. K., Wang, J., Almakhadmeh, Z., & Tolba, A. (2021). Multiple cloud storage mechanism based on blockchain in smart homes. *Future Generation Computer Systems*, *115*, 304–313. DOI: 10.1016/j.future.2020.09.019

Ribot, J. C., Najam, A., & Watson, G. (2009). *Climate Variation, Vulnerability and Sustainable Development in the Semi-arid Tropics*. In Schipper, E. L. F., & Burton, I. (Eds.), *The Earthscan Reader on Adaptation to Climate Change* (pp. 117–160). Earthscan., Available at https://www.researchgate.net/publication/291177311_Climate_variation_vulnerability_and_sustainable_development_in_the_semi-arid_tropics

Rizwan, P., Suresh, K., & Rajasekhara Babu, M. (2017). Real-time smart traffic management system for smart cities by using Internet of Things and big data. *Proceedings of IEEE International Conference on Emerging Technological Trends in Computing, Communications and Electrical Engineering, ICETT 2016*. DOI: 10.1109/ICETT.2016.7873660

Roberts, D. (2008). Thinking globally, acting locally institutionalising climate change at the the local government level in Durban, South Africa. *Environment and Urbanization*, *20*(2), 521–537. DOI: 10.1177/0956247808096126

Rogers, P. P., Jalal, K. F., & Boyd, J. A. (2007). *An Introduction to Sustainable Development*. Harvard University Press., DOI: 10.4324/9781849770477

Rolnick, D., Donti, P. L., Kaack, L. H., Kochanski, K., Lacoste, A., Sankaran, K., & Bengio, Y. (2022). Tackling climate change with machine learning. *Communications of the ACM*, *65*(10), 47–59. DOI: 10.1145/3485128

Ronneberger, O., Fischer, P., & Brox, T. (2015). U-Net: Convolutional Networks for Biomedical Image Segmentation.

Rosa, F., & Bernardino, A. (2019). Advances in Deep Learning for Satellite Image Classification. *Remote Sensing*, *11*(8), 973.

Rother, H. A., Hayward, R. A., Paulson, J. A., Etzel, R. A., Shelton, M., & Theron, L. C. (2022). Impact of extreme weather events on Sub-Saharan African child and adolescent mental health: The implications of a systematic review of sparse research findings . In *Journal of Climate Change and Health* (Vol. 5). DOI: 10.1016/j.joclim.2021.100087

Rout, S. K. (2012). Protection of Plant Varieties and Farmers' Rights in India: A Critical Appraisal. *Journal of Intellectual Property Rights*, *17*, 478–489.

Roy, D. (2020). 'On the Horns of a Dilemma'! Climate change, forest conservation and the marginal people in Indian Sundarbans. *Forum for Development Studies*, 47(2), 307–326. https://doi.org/. 1786452DOI: 10.1080/08039410.2020

Roy, S. (2018). Livelihood resilience of the indigenous Munda community in the Bangladesh Sundarbans Forest. *Handbook Of Climate Change Resilience*, 10(1), 1–21. DOI: 10.1007/978-3-319-71025-9_10-1

Roy, M., Hulme, D., & Jahan, F. (2013). Contrasting adaptation responses by squatters and low-income tenants in Khulna, Bangladesh. *Environment and Urbanization*, *25*(1), 157–176. DOI: 10.1177/0956247813477362

Roy, S. (2019). *Women's Use of the Sundarbans Forest Resources*. Springer Briefs in Environment, Security. *Development and Peace*, *29*, 59–71. DOI: 10.1007/978-981-13-6776-2_4

Rudin, C. (2019). Stop explaining black box machine learning models for high stakes decisions and use interpretable models instead. *Nature Machine Intelligence*, *1*(5), 206–215. DOI: 10.1038/s42256-019-0048-x PMID: 35603010

Rudra, A., & Chattopadhyay, A. (2018). *Environmental vulnerabilities and risks in population: A comparison of two Sundarban Delta*. Population Association of America, May, 1–14. https://www.researchgate.net/publication/326201083

Rudra, A., & Chattopadhyay, A. (2019). Environmental change of coastal Sundarbans: Impact on livelihood and standard of living status of indigenous people. *Environmental Quality Management*, *29*(2), 77–84. DOI: 10.1002/tqem.21667

Russell, S., & Norvig, P. (2021). *Artificial Intelligence: A Modern Approach* (4th ed.). Pearson.

Sahai, S. (2003). India's Plant Variety Protection and Farmers' Rights Act, 2001. *Current Science*, *84*(3), 388–393.

Sahana, M., Rehman, S., Ahmed, R., & Sajjad, H. (2021). Analyzing climate variability and its effects in Sundarban Biosphere Reserve, India: Reaffirmation from local communities. *Environment, Development and Sustainability*, *23*(2), 2465–2492. DOI: 10.1007/s10668-020-00682-5

Saif, S., Russell, A. M., Nodie, S. I., Inskip, C., Lahann, P., Barlow, A., Greenwood Barlow, C., Islam, M. A., & MacMillan, D. C. (2016). Local usage of tiger parts and its role in tiger killing in the Bangladesh Sundarbans. *Human Dimensions of Wildlife*, *21*(2), 95–110. DOI: 10.1080/10871209.2015.1107786

Saito, O., Subramanian, S., Hashimoto, S., & Takeuchi, K. (2020). *Managing socio-ecological production landscapes and seascapes for sustainable communities in Asia mapping and navigating stakeholders, policy and action.* Springer https://www.springer.com/series/11884

Samek, W., Wiegand, T., & Müller, K.-R. (2017). Explainable artificial intelligence: Understanding, visualizing and interpreting deep learning models. *arXiv preprint* arXiv:1708.08296.

Sanchez, T. W., Fu, X., Yigitcanlar, T., & Ye, X. (2024). The Research Landscape of AI in Urban Planning: A Topic Analysis of the Literature with ChatGPT. *Urban Science 2024, Vol. 8, Page 197, 8*(4), 197. DOI: 10.3390/urbansci8040197

Sareen, S., & Haque, M. (2023). Exploring the Evolution and Trends in the Peri-Urban Planning: A Bibliometric Overview. *International Journal of Sustainable Development and Planning, 18*(12), 3855–3866. DOI: 10.18280/ijsdp.181217

Sarkar, S., & Sanyal, P. (2019). Community perception towards tourism development in the Indian Sundarbans: Case study of Jharkhali Island. *International Journal of Engineering Research & Technology (Ahmedabad), 8*(12), 526–540. www.ijert.org

Sarwar Hossain, M., Eigenbrod, F., Amoako Johnson, F., & Dearing, J. A. (2016). Unravelling the interrelationships between ecosystem services and human wellbeing in the Bangladesh delta. *International Journal of Sustainable Development and World Ecology, 24*(2), 120–134. DOI: 10.1080/13504509.2016.1182087

Satpathy, I., Nayak, A., & Jain, V. (2025). The Green City: Sustainable and Smart Urban Living Through Artificial Intelligence. Https://Services.Igi-Global.Com/Resolvedoi/Resolve.Aspx?Doi=10.4018/979-8-3693-6854-1.Ch009, 273–304. DOI: 10.4018/979-8-3693-6854-1.ch009

Satterthwaite, D., Huq, S., Reid, H., Pelling, M., & Romero Lankao, P. (2007). *Adapting to climate change in urban areas: the possibilities and constraints in low- and middle-income nations.* Human Settlements Working Paper, IIED, London. https://www.iied.org/10549iied

Satterthwaite, D., & Mitlin, D. (2014). *Reducing urban poverty in the global south.* Routledge., DOI: 10.4324/9780203104316

Schepers, J., & Wetzels, M. (2007). A meta-analysis of the technology acceptance model: Investigating subjective norm and moderation effects. *Information & Management, 44*(1), 90–103. DOI: 10.1016/j.im.2006.10.007

Schipper, E. L., Ayers, J., Reid, H., Huq, S., & Rahman, A. (Eds.), *Community-based adaptation to climate change: scaling it up* (pp. 214–225). Routledge., DOI: 10.4324/9780203105061

Schmitt, C., & Kwang, H. (2021). A Survey on Artificial Intelligence and Machine Learning in Geospatial Data Processing. *Geo-Spatial Information Science*, *24*(3), 115–131.

Scholz, A. H., Huft, M., & Tvedt, M. W. (2020). Bio-digital prospecting and the dilemma of open science. *Nature Biotechnology*, *38*(4), 389–391. DOI: 10.1038/s41587-020-0482-9

Schwartz, R., Dodge, J., Smith, N. A., & Etzioni, O. (2020). Green AI. *Communications of the ACM*, *63*(12), 54–63. DOI: 10.1145/3381831

Semon, S. F. A. (2003, September). Developments in breeding and techniques within the community plant variety rights fruit sector. In *XI Eucarpia Symposium on Fruit Breeding and Genetics 663* (pp. 707-712).

Sen, H. S. (2019). *The Sundarbans: A disaster-prone ecoregion*. Berlin: Springer. 10, 978-3 https://link.springer.com/content/pdf/10.1007/978-3-030-00680-8.pdf

Seo, S. B., Yadav, P., & Singh, D. (2022). LoRa based architecture for smart town traffic management system. *Multimedia Tools and Applications*, *81*(19), 26593–26608. DOI: 10.1007/s11042-020-10091-5

Serdeczny, O., Adams, S., Baarsch, F., Coumou, D., Robinson, A., Hare, W., Schaeffer, M., Perrette, M., & Reinhardt, J. (2017). Climate change impacts in Sub-Saharan Africa: From physical changes to their social repercussions. *Regional Environmental Change*, *17*(6), 1585–1600. Advance online publication. DOI: 10.1007/s10113-015-0910-2

Settimi, A., Gamerro, J., & Weinand, Y. (2022). Augmented-reality-assisted timber drilling with smart retrofitted tools. *Automation in Construction*, *139*, 104272. DOI: 10.1016/j.autcon.2022.104272

Shafik, W. (2024d). Sustainable development goal 14: explainable AI (XAI) for ocean health. In *Artificial Intelligence and Edge Computing for Sustainable Ocean Health* (pp. 167–198). Springer.

Shafik, W., & Kalinaki, K. (2024). Societal and Ethical Implications of Technology-Enhanced Agriculture and Healthcare: An African Context. *2024 IST-Africa Conference (IST-Africa)*, 1–11. DOI: 10.23919/IST-Africa63983.2024.10569306

Shafik, W. (2024a). Environmental Conservation and Sustainable Development. In *Developing Digital Inclusion Through Globalization and Digitalization* (pp. 142–176). IGI Global. DOI: 10.4018/979-8-3693-4111-7.ch008

Shafik, W. (2024b). Navigating Emerging Challenges in Robotics and Artificial Intelligence in Africa. In *Examining the rapid advance of digital technology in Africa* (pp. 126–146). IGI Global., DOI: 10.4018/978-1-6684-9962-7.ch007

Shafik, W. (2024c). Sustainable Development. In Viana Hassan, A. S. S. B. (Ed.), *Building Community Resiliency and Sustainability With Tourism Development* (1st ed., Vol. 1, pp. 1–30). IGI Global., DOI: 10.4018/979-8-3693-5405-6.ch001

Shah, H. A., Carrasco, L. R., Hamlet, A., & Murray, K. A. (2022). Exploring agricultural land-use and childhood malaria associations in sub-Saharan Africa. *Scientific Reports*, *12*(1), 4124. Advance online publication. DOI: 10.1038/s41598-022-07837-6 PMID: 35260722

Sharifi, A. (2021). Urban sustainability assessment: An overview and bibliometric analysis. *Ecological Indicators*, *121*, 107102. DOI: 10.1016/j.ecolind.2020.107102

Sharma, A., Awasthi, Y., & Kumar, S. (2020). The Role of Blockchain, AI and IoT for Smart Road Traffic Management System. *Proceedings - 2020 IEEE India Council International Subsections Conference, INDISCON 2020*, 289–296. DOI: 10.1109/INDISCON50162.2020.00065

Sharma, D., & Tomar, S. (2010). Mainstreaming climate change adaptation in Indian cities. *Environment and Urbanization*, *22*(2), 451–465. DOI: 10.1177/0956247810377390

Sharma, R., Webster, P., & Bhattacharyya, S. (2014). Factors affecting the performance of community health workers in India: A multi-stakeholder perspective. *Global Health Action*, *7*(1), 25352. DOI: 10.3402/gha.v7.25352 PMID: 25319596

Sheng, S., Song, W., Lian, H., & Ning, L. (2022). Review of urban land management based on bibliometrics. *Land (Basel)*, *11*(11), 1968. DOI: 10.3390/land11111968

Shen, X., & Zhang, C. (2019). A Review on Convolutional Neural Networks for Remote Sensing Image Classification. *International Journal of Applied Earth Observation and Geoinformation*, *79*, 12–23.

Shiva, V. (2000). *Stolen harvest: The hijacking of the global food supply*. South End Press.

Shiva, V. (1997). *Biopiracy: The Plunder of Nature and Knowledge*. South End Press.

Shiva, V. (2016). *Who Really Feeds the World? The Failures of Agribusiness and the Promise of Agroecology*. North Atlantic Books.

Shyamala Devi, M., Suguna, R., Joshi, A. S., & Bagate, R. A. (2019). Design of IoT Blockchain Based Smart Agriculture for Enlightening Safety and Security. *Communications in Computer and Information Science, 985*, 7–19. DOI: 10.1007/978-981-13-8300-7_2

Simatele, D., Binns, T., & Simatele, M. (2012). Sustaining livelihoods under a changing climate: The case of urban agriculture in Lusaka, Zambia. *Journal of Environmental Planning and Management, 55*(9), 1175–1191. Advance online publication. DOI: 10.1080/09640568.2011.637688

Sivaramakrishnan, K. C. (2007). *Democracy in urban India*. Urban Age. https://urbanage.lsecities.net/essays/democracy-in-urban-india

Siyavuş, A. E., & Aydın, T. N. (2022). A bibliometric analysis of urban sprawl. *Papers in Applied Geography, 8*(2), 163–184. DOI: 10.1080/23754931.2021.1975307

Skubis, I., Wolniak, R., & Grebski, W. W. (2024). AI and Human-Centric Approach in Smart Cities Management: Case Studies from Silesian and Lesser Poland Voivodships. *Sustainability 2024, Vol. 16, Page 8279, 16*(18), 8279. DOI: 10.3390/su16188279

Smit, B., & Pilifosova, O. (2003). *From adaptation to adaptive capacity and vulnerability reduction.* In Climate Change, Adaptive Capacity and Development (pp. 9-28). Imperial College Press. DOI: 10.1142/9781860945816_0002

Smit, B., & Skinner, M. W. (2002). Adaptation options in agriculture to climate change: A typology. *Mitigation and Adaptation Strategies for Global Change, 7*(1), 85–114. DOI: 10.1023/A:1015862228270

Smit, B., & Wandel, J. (2006). Adaptation, adaptive capacity and vulnerability. *Global Environmental Change, 16*(3), 282–292. DOI: 10.1016/j.gloenvcha.2006.03.008

Smith, K. (2013). *Environmental hazards: Assessing risk and reducing disaster* (6th ed.). Routledge., DOI: 10.4324/9780203805305

Sokolova, M. V., Fath, B. D., Grande, U., Buonocore, E., & Franzese, P. P. (2024). The role of green infrastructure in providing urban ecosystem services: Insights from a bibliometric perspective. *Land (Basel), 13*(10), 1664. DOI: 10.3390/land13101664

Soliman, M., Cardoso, L., de Almeida, G. F., Araújo, A., & Vila, N. (2021). Mapping smart experiences in tourism: A bibliometric approach. *European Journal of Tourism Research, 28*, 2809–2809. DOI: 10.54055/ejtr.v28i.2254

Son, T. H., Weedon, Z., Yigitcanlar, T., Sanchez, T., Corchado, J. M., & Mehmood, R. (2023). Algorithmic urban planning for smart and sustainable development: Systematic review of the literature. *Sustainable Cities and Society*, *94*, 104562. DOI: 10.1016/j.scs.2023.104562

Sovacool, B. K. (2011). An international comparison of four polycentric approaches to climate and energy governance. *Energy Policy*, *39*(6), 3832–3844. DOI: 10.1016/j.enpol.2011.04.014

Sovacool, B. K. (2018). Bamboo beating bandits: Conflict, inequality, and vulnerability in the political ecology of climate change adaptation in Bangladesh. *World Development*, *102*, 183–194. DOI: 10.1016/j.worlddev.2017.10.014

Spears, N., & Singh, S. N. (2004). Measuring attitude toward the brand and purchase intentions. *Journal of Current Issues and Research in Advertising*, *26*(2), 53–66. DOI: 10.1080/10641734.2004.10505164

StC. (2008). D. E. *Convention on the Conservation of Migratory Species of Wild Animals.*

Stringer, L. C., Fraser, E. D., Harris, D., Lyon, C., Pereira, L., Ward, C. F., & Simelton, E. (2020). Adaptation and development pathways for different types of farmers. *Environmental Science & Policy*, *104*, 174–189. DOI: 10.1016/j.envsci.2019.10.007

Strubell, E., Ganesh, A., & McCallum, A. (2019). Energy and policy considerations for deep learning in NLP. *Proceedings of the 57th Annual Meeting of the Association for Computational Linguistics*, 3645–3650. DOI: 10.18653/v1/P19-1355

Subramanian, V., Holzer, A., & Weis, T. M. (2021). Building resilient AI: Principles and challenges. *AI and Ethics*, *1*(1), 43–53. DOI: 10.1007/s43681-020-00003-1

Sundarbans National park. (1987). *UNESCO World Heritage Centre*. https://whc.unesco.org/en/list/452

Szabo, S., Hossain, M. S., Adger, W. N., Matthews, Z., Ahmed, S., Lázár, A. N., & Ahmad, S. (2016). Soil salinity, household wealth and food insecurity in tropical deltas: Evidence from south-westcoast of Bangladesh. *Sustainability Science*, *11*(3), 411–421. DOI: 10.1007/s11625-015-0337-1 PMID: 30174734

Tan, J., Yuan, Q., Guo, W., Xie, N., Liu, F., Wei, J., & Zhang, X. (2022). Deep Reinforcement Learning for Traffic Signal Control Model and Adaptation Study. *Sensors 2022, Vol. 22, Page 8732*, *22*(22), 8732. DOI: 10.3390/s22228732

Tan, B. C., & Lau, T. C. (2011). Green purchase behavior: Examining the influence of green environmental attitude, perceived consumer effectiveness and specific green purchase attitude. *Australian Journal of Basic and Applied Sciences*, *5*(8), 559–567.

Tao, M., Qu, C., Wei, W., Zhou, B., & Huang, S. (2018). Hybrid Cloud Architecture for Cross-Platform Interoperability in Smart Homes. *Lecture Notes in Computer Science (Including Subseries Lecture Notes in Artificial Intelligence and Lecture Notes in Bioinformatics), 11336 LNCS*, 608–617. DOI: 10.1007/978-3-030-05057-3_45

Thakur, S., Maity, D., Mondal, I., Basumatary, G., Ghosh, P. B., Das, P., & De, T. K. (2021). Assessment of changes in land use, land cover, and land surface temperature in the mangrove forest of Sundarbans, northeast coast of India. *Environment, Development and Sustainability*, *23*(2), 1917–1943. DOI: 10.1007/s10668-020-00656-7

Tiepolo, M., & Galligari, A. (2021). Urban expansion-flood damage nexus: Evidence from the Dosso Region, Niger. *Land Use Policy*, *108*, 105547. Advance online publication. DOI: 10.1016/j.landusepol.2021.105547

Tonapi, V. A. (2008). PPV and FR Act and National Seed Policy. *Sorghum Improvement in the New Millennium. Patancheru 502 324, Andhra Pradesh, India: International Crops Research Institute for the Semi-Arid Tropics. 340 pp*, *48*, 232.

Trafimow, D., & Fishbein, M. (1994). The moderating effect of behavior type on the subjective norm-behavior relationship. *The Journal of Social Psychology*, *134*(6), 755–763. DOI: 10.1080/00224545.1994.9923010

Trafimow, D., Sheeran, P., Conner, M., & Finlay, K. A. (2002). Evidence that perceived behavioural control is a multidimensional construct: Perceived control and perceived difficulty. *British Journal of Social Psychology*, *41*(1), 101–121. DOI: 10.1348/014466602165081 PMID: 11970777

Tran, T. V., Dang, N. T., & Chung, W. Y. (2017). Battery-free smart-sensor system for real-time indoor air quality monitoring. *Sensors and Actuators. B, Chemical*, *248*, 930–939. DOI: 10.1016/j.snb.2017.01.198

Tripodi, T., & Potocky-Tripodi, M. (2007). *International Social Work Research: Issues and Prospects*. Oxford University Press., DOI: 10.1093/acprof:oso/9780195187250.001.0001

Twigg, J. (2007). *Characteristics of a disaster-resilient community, a guidance note*. Available at https://www.researchgate.net/publication/315799865_Characteristics_of_a_Disaster-resilient_Community_A_Guidance_Note_Characteristics_of_a_Disaster-resilient_Community_A_Guidance_Note

Tyagi, N., & Bhushan, B. (2023). Demystifying the Role of Natural Language Processing (NLP) in Smart City Applications: Background, Motivation, Recent Advances, and Future Research Directions. *Wireless Personal Communications*, *130*(2), 857–908. DOI: 10.1007/s11277-023-10312-8 PMID: 37168438

Tyagi, S. K. S., Mukherjee, A., Pokhrel, S. R., & Hiran, K. K. (2020). An intelligent and optimal resource allocation approach in sensor networks for smart agri-IoT. *IEEE Sensors Journal*, *21*(16), 17439–17446. DOI: 10.1109/JSEN.2020.3020889

Tyler, S., & Moench, M. (2012). A framework for urban climate resilience. *Climate and Development*, *4*(4), 311–326. DOI: 10.1080/17565529.2012.745389

Uddin, S. M. (2019). Religion, Nature, and Life in the Sundarbans. *Asian Ethnology*, *78*(2), 289–310. https://www.jstor.org/stable/26845329

UNDRR. (2015). Sendai Framework for Disaster Risk Reduction 2015–2030. United Nations Office for Disaster Risk Reduction. https://www.undrr.org/publication/sendai-framework-disaster-risk-reduction-2015-2030

UNESCO. (2021). *Recommendation on the Ethics of Artificial Intelligence.* Retrieved from https://unesdoc.unesco.org/ark:/48223/pf0000381137

United Nations Office for Disaster Risk Reduction (UNDRR). (2015). *Sendai Framework for Disaster Risk Reduction 2015–2030.* https://www.undrr.org/publication/sendai-framework-disaster-risk-reduction-2015-2030

United Nations Office for Disaster Risk Reduction (UNDRR). (2019). *Global Assessment Report on Disaster Risk Reduction.* https://gar.undrr.org

United Nations. (2018). *United Nations Declaration on the Rights of Peasants and Other People Working in Rural Areas (UNDROP).* https://www.ohchr.org

Unogwu, O. J., Doshi, R., Hiran, K. K., Mijwil, M. M., Catherine, A. T., & Abotaleb, M. (2024). *Exploring the Implications of Emerging Artificial Intelligence Technologies at Edge Computing in Higher Education.* In Integrating Generative AI in Education to Achieve Sustainable Development Goals (pp. 169-182). IGI Global. DOI: DOI: 10.4018/979-8-3693-2440-0.ch009

Uttajug, A., Ueda, K., Seposo, X., & Francis, J. M. (2023). Association between extreme rainfall and acute respiratory infection among children under-5 years in sub-Saharan Africa: An analysis of Demographic and Health Survey data, 2006-2020. *BMJ Open*, *13*(4), e071874. Advance online publication. DOI: 10.1136/bmjopen-2023-071874 PMID: 37185183

Van Winkle, T., Kotval-K, Z., Machemer, P., & Kotval, Z. (2022). Health and the urban environment: A bibliometric mapping of knowledge structure and trends. *Sustainability (Basel), 14*(19), 12320. DOI: 10.3390/su141912320

Vangala, A., Das, A. K., Kumar, N., & Alazab, M. (2020). Smart Secure Sensing for IoT-Based Agriculture: Blockchain Perspective. *IEEE Sensors Journal, 21*(16), 17591–17607. DOI: 10.1109/JSEN.2020.3012294

Velasco, A. C. (2020). The CJEU reaches a controversial compromise on the scope of protection of plant variety rights. *Journal of Intellectual Property Law & Practice, 15*(12), 976–983. DOI: 10.1093/jiplp/jpaa159

Venkatesan, R. (2018). TRIPS and Plant Variety Protection in India: Complicating the Globalisation Debate. *Indian J. Int'l Econ. L., 9*, 43.

Verma, S., Gala, R., Madhavan, S., Burkule, S., Chauhan, S., & Prakash, C. (2018). An Internet of Things (IoT) Architecture for Smart Agriculture. *Proceedings - 2018 4th International Conference on Computing, Communication Control and Automation, ICCUBEA 2018*. DOI: 10.1109/ICCUBEA.2018.8697707

Verma, D. D. P. (2010). Review of Article 27.3 (B) Under TRIPs Agreement: A Critical Analysis. *Available atSSRN* 1587584. DOI: 10.2139/ssrn.1587584

Vinuesa, R., Azizpour, H., Leite, I., Balaam, M., Dignum, V., Domisch, S., & Nerini, F. F. (2020). The role of artificial intelligence in achieving the Sustainable Development Goals. *Nature Communications, 11*(1), 233. DOI: 10.1038/s41467-019-14108-y PMID: 31932590

Visser, B., Eaton, D., Louwaars, N., & van Dusen, M. E. (2005). *Transaction costs of germplasm exchange under bilateral agreements*. Global Forum on Agricultural Research.

Wagner, N. (2019). Sustainability in Port cities–a bibliometric approach. *Transportation Research Procedia, 39*, 587–596. DOI: 10.1016/j.trpro.2019.06.060

Wallace, R. B., Horsfall, F., Goubran, R., El-Haraki, A., & Knoefel, F. (2019, May 3). The Challenges of Connecting Smart Home Health Sensors to Cloud Analytics. *SAS 2019 - 2019 IEEE Sensors Applications Symposium,Conference Proceedings*. DOI: 10.1109/SAS.2019.8706128

Wang, C., Lv, T., Cai, R., Xu, J., & Wang, L. (2022). Bibliometric analysis of multi-level perspective on sustainability transition research. *Sustainability (Basel), 14*(7), 4145. DOI: 10.3390/su14074145

Wang, H., Pan, Y., & Luo, X. (2019). Integration of BIM and GIS in sustainable built environment: A review and bibliometric analysis. *Automation in Construction, 103*, 41–52. DOI: 10.1016/j.autcon.2019.03.005

Wang, H., Xu, P., & Zhao, J. (2021). Improved KNN Algorithm Based on Preprocessing of Center in Smart Cities. *Complexity, 2021*(1), 5524388. DOI: 10.1155/2021/5524388

Wang, M. H., Ho, Y. S., & Fu, H. Z. (2019). Global performance and development on sustainable city based on natural science and social science research: A bibliometric analysis. *The Science of the Total Environment, 666*, 1245–1254. DOI: 10.1016/j.scitotenv.2019.02.139 PMID: 30970489

Wezel, A., Bellon, S., Doré, T., Francis, C., Vallod, D., & David, C. (2009). Agroecology as a science, a movement and a practice. *Sustainable Agriculture, 29*(4), 503–515.

Wilkinson, M. D., Dumontier, M., Aalbersberg, I. J., Appleton, G., Axton, M., Baak, A., Blomberg, N., Boiten, J.-W., da Silva Santos, L. B., Bourne, P. E., Bouwman, J., Brookes, A. J., Clark, T., Crosas, M., Dillo, I., Dumon, O., Edmunds, S., Evelo, C. T., Finkers, R., & Mons, B. (2016). The FAIR Guiding Principles for scientific data management and stewardship. *Scientific Data, 3*(1), 160018. DOI: 10.1038/sdata.2016.18 PMID: 26978244

Winfield, A. F., & Jirotka, M. (2018). Ethical governance is essential to building trust in robotics and artificial intelligence systems. *Philosophical Transactions. Series A, Mathematical, Physical, and Engineering Sciences, 376*(2133), 20180085. DOI: 10.1098/rsta.2018.0085 PMID: 30323000

Winge, T., Andersen, R., & Ramanna-Pathak, A. (2013). Combining Farmers' Rights and plant variety protection in Indian law. In *Realising Farmers' Rights to Crop Genetic Resources* (pp. 54-61). Routledge.

World Bank Group. (2025). Agriculture and Food. https://www.worldbank.org/en/topic/agriculture/overview

WTO. (1994). *The Agreement on Trade-Related Aspects of Intellectual Property Rights (TRIPS).*

Wynberg, R., & Chennells, R. (2009). Green Governance: An Institutional Analysis of ABS and Traditional Knowledge. In Wynberg, R., Schroeder, D., & Chennells, R. (Eds.), *Indigenous Peoples, Consent and Benefit Sharing* (pp. 189–206). Springer. DOI: 10.1007/978-90-481-3123-5

Xiong, W., Wang, Y., Gao, S., Huang, X., & Wang, S. (2024). A multi-objective service composition optimization method considering multi-user benefit and adaptive resource partitioning in hybrid cloud manufacturing. *Journal of Industrial Information Integration, 38*, 100564. DOI: 10.1016/j.jii.2024.100564

Yang, G., Zhang, P., Yu, F., & Zhu, X. (2023). A review on resilient cities research from the perspective of territorial spatial planning: A bibliometric analysis. *Frontiers in Ecology and Evolution, 11*, 1300764. DOI: 10.3389/fevo.2023.1300764

Yang, J., & Wang, C. (2017). Deep Learning for Geospatial Image Analysis: Applications and Challenges. *Remote Sensing, 9*(3), 281.

Yang, Q., Liu, Y., Chen, T., & Tong, Y. (2019). Federated machine learning: Concept and applications. *ACM Transactions on Intelligent Systems and Technology, 10*(2), 12. DOI: 10.1145/3298981

Yao, H., Gao, P., Wang, J., Zhang, P., Jiang, C., & Han, Z. (2019). Capsule Network Assisted IoT Traffic Classification Mechanism for Smart Cities. *IEEE Internet of Things Journal, 6*(5), 7515–7525. DOI: 10.1109/JIOT.2019.2901348

Yao, X., Yang, Y., & Lin, H. (2018). Geospatial Data Analytics Using Machine Learning Techniques: A Survey. *Geospatial Intelligence Review, 17*(2), 90–112.

Yassine, A., Singh, S., Hossain, M. S., & Muhammad, G. (2019). IoT big data analytics for smart homes with fog and cloud computing. *Future Generation Computer Systems, 91*, 563–573. DOI: 10.1016/j.future.2018.08.040

Yilmaz, O. N. C., Li, Z., Valkealahti, K., Uusitalo, M. A., Moisio, M., Lundén, P., & Wijting, C. (2014). Smart mobility management for D2D communications in 5G networks. *2014 IEEE Wireless Communications and Networking Conference Workshops. WCNCW, 219–223*, 219–223. Advance online publication. DOI: 10.1109/WCNCW.2014.6934889

Yohe, G., & Tol, R. S. J. (2002). Indicators for social and economic coping capacity moving toward a working definition of adaptive capacity. *Global Environmental Change, 12*(1), 25–40. DOI: 10.1016/S0959-3780(01)00026-7

Yu, J., Guo, L., Zhang, J., & Wang, G. (2024). A survey on graph neural network-based next POI recommendation for smart cities. *Journal of Reliable Intelligent Environments, 10*(3), 299–318. DOI: 10.1007/s40860-024-00233-z

Zaidi, A., Ajibade, S. S. M., Musa, M., & Bekun, F. V. (2023). New insights into the research landscape on the application of artificial intelligence in sustainable smart cities: A bibliometric mapping and network analysis approach. *International Journal of Energy Economics and Policy, 13*(4), 287–299. DOI: 10.32479/ijeep.14683

Zhang, H., & Li, W. (2018). An Overview of Geospatial Big Data and AI Technologies. *Geospatial Data Science*, *42*(4), 58–75.

Zhang, L., & Zhang, L. (2020). Deep Learning for Geospatial Data Analysis: A Review. *Geospatial Data Science*, *42*(3), 45–58.

Zheng, Y., Lin, Y., Zhao, L., Wu, T., Jin, D., & Li, Y. (2023). Spatial planning of urban communities via deep reinforcement learning. *Nature Computational Science 2023 3:9, 3*(9), 748–762. DOI: 10.1038/s43588-023-00503-5

Zhou, S., Wei, C., Song, C., Pan, X., Chang, W., & Yang, L. (2023). Short-Term Traffic Flow Prediction of the Smart City Using 5G Internet of Vehicles Based on Edge Computing. *IEEE Transactions on Intelligent Transportation Systems*, *24*(2), 2229–2238. DOI: 10.1109/TITS.2022.3147845

Zhu, X., & Wang, H. (2020). AI for Geospatial Intelligence: Applications and Challenges. *Geospatial Intelligence Journal*, *27*(3), 231–245.

About the Contributors

Walid Chouari is an Associate Professor in Geography at the College of Arts, King Faisal University, Saudi Arabia. His research interests include Mapping of flood risk and flood-prone areas, Mapping of the dynamics of wet depressions and Geostatistical and environmental mapping.

Abdelkader Mohamed Sghaier Derbali is an Assistant Professor in Finance at the Department of Finance and Accounting, Higher Institute of Informatics and Management of Kairouan, Kairouan University, Kairouan, Tunisia and at the Department of Administrative and Financial Sciences, Applied College, Taibah University, Saudi Arabia. He is one of the Editorial Board Members in the Journal of Energy Markets, International Business Review, Cogent Economic & Finance, International Review of Applied Economics, African Journal of Accounting, Auditing and Finance, and African Journal of Economic and Sustainable Development. His research interests include Risk Management, Systemic risk, International finance, Capital markets and institutions, Banking and market microstructure and Islamic Finance. He has published articles, among others, in Research in International Business and Management, Cogent Business & Management, Cogent Economic & Finance, Journal of Chinese Governance, The Chinese Economy, Journal of Energy Markets, and International Journal of Trade and Global Markets.

***

Oindrilla Ghosh is a Research Scholar hailing from the School of Law at Christ (Deemed to be University) Bangalore, India, who focuses her research on the health of tribal communities. Her work includes Neuroscience on Organizational Behavior publications at IGI Global and Environmental Disasters, of which 8 of them are indexed in SCOPUS and Web of Sciences conference proceedings. She completed her early education at Loreto Day School, Bowbazar, and achieved distinction in her undergraduate (B.A/LL.B) and postgraduate (LL.M) studies at Calcutta University. Having served as an Assistant Professor of Law at Kingston Law College, Barasat, Kolkata, Ghosh has actively participated in numerous National Conferences, International Conferences, UN Global Summits, webinars, and seminars organized by prestigious academic institutions. Her expertise lies in health perspectives, environmental studies, human behavior, and related areas. Additionally, she has conducted various workshops and lectures in educational institutions, and is engaged with NGO's to help the needy

Sharath Shridhar Hegde is a researcher in the Department of Management Studies, Dayananda Sagar College of Engineering, Bangalore, India. Has interest in the area of Marketing, Fiancne and Agriculture.

Ganesha K S is an Assistant Professor in the Department of Management Studies, at Dayananda Sagar College of Engineering, Bangalore, India. He has 14 years of experience in Corporate, Teaching and Research. He is a Member of the International Safety Quality Environment Management Association (ISQEM), UK. His primary research interests are tourist behavior, social media, sentiment analysis, human resource management, and labor studies. He has presented at multiple conferences and published articles in Scopus and other International Indexed journals. He has conducted FDP, workshops, and training in the areas of Digital Marketing, Statistical Analysis using SPSS, Project Management, Research Methodology, Entrepreneurship etc.

Achyutananda Mishra is a Professor at School of Law, Christ University, Bangalore. He has 25 years of teaching experience. He has his research interests in the areas of Constitutional law, Legal theory, Democracy and Globalization. He has supervised a Postdoctoral research scholar and is currently guiding Ph.D. scholars under his supervision.

Ananya Pandey is a Research Scholar in School of Law at Christ University, Bangalore. She has her research interests in the field of Constitutional Law, Environment Law, Human Rights, Food Security, Intellectual Property Law and Jurisprudence.

Wasswa Shafik (Member, IEEE) received a Bachelor of Science degree in information technology from Ndejje University, Uganda, a Master of Engineering degree in Information Technology (Computer and Communication Networks) from Yazd University, Iran, and a PhD in Computer Science with the School of Digital Science, Universiti Brunei Darussalam, Brunei Darussalam. He is also the Founder and Research Director of the Dig Connectivity Research Laboratory (DCRLab) after serving as a Research Associate at the Network Interconnectivity Research Laboratory at Yazd University. Prior to this, he worked as a Community Data Analyst at Population Services International (PSI-Uganda), a Community Data Officer at the Programme for Accessible Health Communication (PACE-Uganda), a Research Assistant at the Socio-Economic Data Centre (SEDC-Uganda), Uganda, an Assistant Data Officer at TechnoServe, Kampala, Uganda, and Asmaah Charity Organization as Cheif Executive Officer. He has hundreds of publications with renowned publishers. His research interests include Artificial Intelligence, Neural Networks, Computer Vision, Smart Agriculture, Digital Health, and Green Computing.

Rohit Sood, Assistant Professor at Lovely Professional University, Phagwara, Punjab (India), Department of Finance, has an extensive academic and professional experience of more than 14 years. He has presented numerous papers in national and international conferences and ten papers are published in reputed international referred journals with high impact factor and indexing Google Scholar, Scopus, BEIESP, Open J Gate, DOAJ, get CITED, dig, Higher Education Forum, Cite Factor, Index Copernicus, Search Sight, Scirus, UK Index, New Jour, Research Gate. Apart from this, he has attended various conferences, workshop and seminar in different colleges. He has attended 10+ relevant training and courses for continuous learning; has attended more than 10 Faculty Development Programs. He is a life member of Society of FSP and Global Risk Community.

Manjeri Subin Sunder Raj, a fifth-generation lawyer, enrolled in 2007 and is extremely passionate about Rights of Nature, Environmental Rights and Human Rights. He is involved in Public Interest Litigations and several initiatives involving coordination and conduction of environmental protection campaigns, animal and human rights issues, rural extension programmes and legal aid clinics, in India. He is the first and only Indian to be a part of the Expert Member Group of the Harmony with Nature Network, United Nations. Dr. Subin was a Linnaeus-Palme Fellow at the Royal Institute of Technology (KTH - Kungliga Tekniska högskolan), Stockholm, Sweden, in 2014 and 2015 and delivered lectures on Water Law. He was one among the 15 selected from around the world, and invited to present his doctoral work at the First Postgraduate Colloquium on the Frontiers of International Environmental Law held at the Faculty of Law, University of Oslo, in September 2017. His doctoral thesis, Earth Justice: Developing a New Jurisprudence has been published by Thomson Reuters in 2019. He is the co-author of Enforcing Hazardous Wastes Rules in India: Strategies and Techniques for Achieving Increased Compliance (Environmental Law Institute, Washington D.C, 2014) and a contributing author of Water and Scriptures (Springer International Publishing, 2017). He strives to ensure that there is a marked change in human attitude by focusing more on the 'right of environment'. This, he believes, is to be achieved by imposing affirmative duties, on man, by harnessing a value-based rationale which fosters environmental stewardship. He believes that realizing intrinsic value is paramount, and in instilling in man the principles of Earth Justice. One way to achieve this, he believes, is to afford a legal personality to non-human entities, with a special thrust on animals, so as to ensure that their intrinsic rights are protected. Dr. Subin completed his graduation from the National University of Advanced Legal Studies (NUALS) securing a second rank, in 2007. He did his LLM in 2011 and his PhD from the National Law School of India University, Bengaluru, India, in 2018.

Index

T

U

W

www.ingramcontent.com/pod-product-compliance
Lightning Source LLC
LaVergne TN
LVHW081333300926
849161LV00008B/1957
* 9 7 9 8 3 3 7 3 3 2 0 6 2 *